●大学英语拓展课程“十三五”规划教材●

中国民居建筑英语

English for the Chinese Traditional Vernacular Architecture

◎主　编　黄运亭　索全兵

◎副主编　余潇潇　王亚冰

华南理工大学出版社
SOUTH CHINA UNIVERSITY OF TECHNOLOGY PRESS
·广州·

图书在版编目（CIP）数据

中国民居建筑英语/黄运亭，索全兵主编．—广州：华南理工大学出版社，2019. 5
（大学英语拓展课程“十三五”规划教材）
ISBN 978 -7 -5623 -5309 -6

Ⅰ. ①中…　Ⅱ. ①黄…②索…　Ⅲ. ①居民 - 建筑设计 - 英语 - 高等学校 - 教材
Ⅳ. ①TU241. 5

中国版本图书馆 CIP 数据核字（2017）第 290861 号

中国民居建筑英语
黄运亭　索全兵　主编

出 版 人：卢家明
出版发行：华南理工大学出版社
（广州五山华南理工大学 17 号楼，邮编 510640）
http：//www. scutpress. com. cn　E-mail：scutc13@ scut. edu. cn
营销部电话：020 - 87113487　87111048（传真）
总 策 划：乔　丽
策划编辑：吴翠微
责任编辑：吴翠微　陈　蓉
印 刷 者：虎彩印艺股份有限公司
开　　本：787mm × 1092mm　1/16　**印张：**9. 5　**字数：**246 千
版　　次：2019 年 5 月第 1 版　2019 年 5 月第 1 次印刷
定　　价：29. 00 元

前　言

在经济全球化、科学技术一体化的时代，作为当今世界使用最为广泛的信息载体和交流工具，英语发挥着越来越重要的作用。随着我国对外交流日益频繁，国家和社会对于大学英语教学提出了更高的要求，《中国民居建筑英语》就是为了满足建筑专业大学英语教学需要而编写的。

本教材以 ESP 教育理念为基础，以“任务型教学”训练为目标，以中国传统建筑文化和中国各地传统代表性建筑为主要内容，着重训练学生的阅读、翻译和口语能力，以期给建筑类专业的 ESP 教学提供一本有用的教材。

本教材由 10 个单元构成。前 3 个单元介绍中国传统建筑文化和特点，后 7 个单元分别介绍国内具有代表性的传统民居建筑及其特点，并配有大量图片。

本教材的练习体现了“任务驱动”的理念，强调以学生为中心，以职业活动为导向，以任务为载体，让学生在做中学、学中做，在任务的完成中构建应用语言的综合能力。每单元的练习紧扣该单元主题，安排合理，形式多样，以知识的输入促进语言的输出，同时为学生提供英语学习的拓展空间，激发学生自主学习。

本教材选材内容新颖，素材详实，图文并茂，可作为建筑专业英语教材，也可供建筑设计人员参考。

华南理工大学刘涛波教授对书稿进行了审校，在此，本书编者谨对刘教授的无私奉献和辛勤付出表示崇高的敬意和衷心的感谢。

由于编者水平有限，错误在所难免，恳请广大读者和同行专家批评指正。

编　者

2019 年 3 月

Contents

Unit 1

The History and Culture of the Chinese Traditional Vernacular Architecture

▶ Pre-reading Activities

1. Do you know the layout of a modern house? Discuss in small groups about what your home looks like. You may use the following words for reference: roof, basement, living room, bedroom, dining room, kitchen, bathroom, corridor, hallway, courtyard, lawn, garage.
2. What do you know about ancient residences? How were they different from modern ones?
3. Look at the following pictures and tell what they are and when they were used. Then skim the text to see if you are right.

Text A

The Origin and Development of the Chinese Traditional Vernacular Architecture

Clothes, food, accommodation and transportation are the four basic necessities of man's life. To meet the needs of accommodation, housing architecture came into being and became the earliest architecture type in the history.

In the ancient times, the primitive people dwelled in natural caves, which was referred to as cave-living. In the South, due to the hot and humid climate, insects and wild animals in the open country, the crude shanty in the trees became the habitat, known as the nesting. Later, people began to build their habitat half underground, and then on the ground. A large number of these earliest residential sites have been found in Banpo of Shaanxi Province in China.

In the lower reaches of the Yangtze River, ruins of a village were excavated in Hemudu, Yuyao, Zhejiang Province. It is a long bole-fence house stretching more than 30 meters, built on wood piles with the spatial depth of around 7 meters and front brim of 1. 3 meters wide. The house floor is about 1 meter higher than the ground, with a wooden ladder for access. The tenon and mortise joints connecting other components such as beams have a complex structure, indicating that there was a big step forward in the wood construction technology at that time.

The feudal society gave priority to agricultural production at its early stage. From the unearthed painting stones and bricks, afterlife utensils and pottery houses from the tombs, we can infer that the small houses in the Han Dynasty were usually of square or rectangle shape. Relatively larger houses were designed in tri-square shape, having a courtyard inside, with the high building in the middle and sub-houses being relatively low so that the contour of the residence has a clear distinction between the major and the minor houses. A large-scale aristocratic residence is composed of several courtyards and besides the main building are rooms for miscellaneous purposes and guests. Inside the front hall is the main building and rooms are placed in the rear hall, which is developed from the ancient front-hall-rear-room (前堂后室) principle.

The stone carvings in the Northern and Eastern Wei periods show that the aristocratic houses are built with large halls and corridors, with even gardens in the rear of the residence. It is the prototype of the garden house.

The vernacular architectures of the Sui and Tang Dynasties and the Five Dynasties can only be found in the Dunhuang mural paintings and some other paintings. For instance, the entrance door of an aristocratic mansion adopts tortoise-head shape; two major buildings inside the residence are connected with cloister with straight frame windows, forming a courtyard.

Rural dwellings in the Song Dynasty can be found in the famous painting of that time, *Along the River During the Qingming Festival*, thatching huts being relatively simple and primitive, with low walls, and a group of cottages and tile houses. Small houses in cities and towns are of rectangular shape, having overhanging gable roof or *xieshan*-style roof (roof with four sloping surfaces and nine hips). In addition to grass and tile coverage, two sides of the roof and front eaves use bamboo coverage or skylight roof is adopted. For a larger house, a door-house is built outside and the inside is built in the form of courtyard with trees and flowers planted for landscaping.

Besides, in his famous painting of *A Thousand Li of Rivers and Mountains*, Wang Ximeng of the Song Dynasty painted some houses and each has an entrance door and chambers in eastern and western wings. Yet, the main part of the residence is a 工-shape house consisting of the front hall, corridor and rooms at rear. The 工-shape or 王-shape plane connected by a corridor in the middle, which was

built in accordance with the traditional layout of a front-hall-rear-room in ancient time, became a layout feature of the vernacular house in that period.

During the period of Ming and Qing Dynasties, a surge in population, economic prosperity and cultural development brought about the prosperity in urban and rural areas. Considerable vernacular architectures all over the nation, retaining the traditional characteristics of various ethnic groups in different regions, have survived to this day.

New Words and Expressions

vernacular /vərˈnækjələr/	*adj.*	民间风格的
primitive /ˈprɪmətɪv/	*adj.*	belonging to a society in which people live in a very simple way 原始的，远古的；粗糙的
dwell /dwel/	*v.*	inhabit or live in 居住
shanty /ʃænti/	*n.*	small crude shelter used as a dwelling 棚屋
habitat /ˈhæbɪtæt/	*n.*	the type of environment in which an organism or group normally lives or occurs 栖息地
nesting /nestɪŋ/	*n.*	a structure in which an organism lays eggs or gives birth to their young 巢穴
ruin /ˈruːɪn/	*n.*	an irrecoverable state of devastation and destruction 废墟，遗址
excavate /ˈekskəveɪt/	*v.*	find by digging in the ground 发掘
bole-fence	*n.*	树干篱笆
spatial /ˈspeɪʃl/	*adj.*	pertaining to or involving or having the nature of space 空间的
brim /brɪm/	*n.*	the top edge of a vessel or other container 边，边缘
tenon /ˈtenən/	*n.*	a projection at the end of a piece of wood that is shaped to fit into a mortise 榫
mortise /ˈmɔːrtɪs/	*n.*	a square hole made to receive a tenon and so to form a joint 榫眼
beam /biːm/	*n.*	long thick piece of wood or metal or concrete, etc., used in construction 梁，栋梁
feudal /ˈfjuːdl/	*adj.*	封建制度的
unearth /ʌnˈɜːrθ/	*v.*	lay bare through digging 挖掘，发掘
utensil /juːˈtensl/	*n.*	an implement for practical use (especially in a household) 用具，器皿

pottery /ˈpɑːtəri/	*n.*	earthenware pots, etc., made by hand 陶器
tri-square	*n.*	tool used for marking and measuring a piece of wood 曲尺
courtyard /ˈkɔːrtjɑːrd/	*n.*	an area wholly or partly surrounded by walls or buildings 庭院
contour /ˈkɑːntʊr/	*n.*	the general shape or outline of an object 轮廓
aristocratic /əˌrɪstəˈkrætɪk/	*adj.*	贵族的
miscellaneous /ˌmɪsəˈleɪniəs/	*adj.*	consisting of many different kinds of things or people that are difficult to put into a particular category 各式各样的
rear /rɪr/	*n.*	the side of an object that is opposite its front 后面
corridor /ˈkɔːrədər/	*n.*	a long passage in a building or train, with doors and rooms on one or both sides 走廊，过道
prototype /ˈproʊtətaɪp/	*n.*	a standard or typical example 原型，模型
mansion /ˈmænʃn/	*n.*	a very large house 大厦
thatch /θætʃ/	*v.*	用茅草盖屋顶
hut /hʌt/	*n.*	a small house with only one or two rooms, especially one which is made of wood, mud, grass, or stones 小屋，棚屋
tile /taɪl/	*n.*	pieces of baked clay, carpet, cork, or other substance, which are fixed as a covering onto a floor or wall 瓦片，瓷砖
eaves /iːvz/	*n.*	the edges of a roof that project beyond the wall 屋檐
layout /ˈleɪaʊt/	*n.*	the way in which the parts of a building, a garden or something else are arranged 布局
ethnic /ˈeθnɪk/	*adj.*	belonging to or deriving from the cultural, religious, or linguistic traditions of people or a country 种族的

Exercises

Task 1 Discovering the Main Ideas

1. Answer the following questions with the information contained in Text A.

1) What are the four necessities of man's life?

2) Why did primitive people live on trees?

3) What was the general shape of the small houses in the Han Dynasty?
4) What are the general differences between the houses in the Sui Dynasty and those in the Song Dynasty?

2. Text A can be divided into three parts with the paragraph number(s) of each part provided as follows. Write down the main idea of each part.

Part	Paragraph(s)	Main idea
One	1 – 1	______________________ ______________________
Two	2 – 3	______________________ ______________________
Three	4 – 9	______________________ ______________________

Task 2 Reading Between Lines

3. Match each housing architecture with the period it first appeared.

courtyard	Hemudu people
cave living	Northern and Eastern Wei periods
garden house	Sui and Tang Dynasties
bole-fence house	Song Dynasty
工-shape house	Primitive people

4. Fill in the blanks with the given words. You may not use any of the words in the bank more than once. Change the form of the given words if necessary.

accommodation	dwell	habitat	excavate	stretch	access
priority	distinction	rear	prototype	surge	characteristics

1) We ______ in the country but work in the city.
2) They plan to ______ a large hole before putting in the foundation.
3) Regardless of where we are and what we are doing, we want ______ to our data.
4) Many species are in peril of extinction because of our destruction of their natural ______.
5) At a time of this economic crisis, our ______ should be very clear about what we need to do.
6) There is no neat ______ between operating system software and the software that runs

on top of it.

7) Parents transmit some of their ______ to their children.

8) Many people in America and Europe think that the recent ______ in inflation, like almost everything else these days, is "made in China".

9) The company showed the ______ of the new model at the exhibition.

10) I lie still for a minute until I can breathe normally, and then ______ my arms out to prop myself up.

Task 3 Challenge Yourself

5. Translate the following paragraph into Chinese.

During the period of Ming and Qing Dynasties, a surge in population, economic prosperity and cultural development brought about the prosperity in urban and rural areas. Considerable vernacular architectures all over the nation, retaining the traditional characteristics of various ethnic groups in different regions, have survived to this day.

Text B

The Chinese Traditional Vernacular Architecture and Culture

China has a long history, vast land and rich cultural heritage. As the most substantial architectural culture heritages are closely related to the production and daily life of the ordinary people, the traditional vernacular house is equally abundant. Today, it is spreading in different ethnic groups and regions all over the nation. After the vicissitudes of the time, it is still widely used by the ordinary people and some have shown us the quintessence of the fine tradition and exquisite art. It is a precious cultural treasure of China.

Apart from the disparities in climate in north and south China, different geographical conditions, material resources, customs, lifestyles and aesthetic requirements of various ethnic groups have resulted in the distinctive ethnic features and rich local characteristics in the Chinese traditional vernacular architecture.

The Chinese traditional vernacular architecture ties up with the society, history, culture, nationality and folk-customs, as well as the theories of the Confucian rites, Taoism, *Yin-Yang* and Five Elements. The outstanding traditional vernacular architecture is of historic and cultural value and of practical and artistic value as well. It is not only the precious cultural heritage of China, but also the precious cultural wealth of the world, calling for urgent protection and promotion.

Philosophy is a concentrated expression of culture, the quintessence of the national spirit and the acme of human wisdom. In the system of the traditional Chinese culture, philosophy is at the core position, playing a leading role.

From a philosophical perspective, the traditional Chinese culture is influenced by the following ideologies, namely, Confucianism, Taoism, the Theory of *Yin-Yang* and Five Elements and Chinese Buddhism in later time, which have absorbed, impacted and integrated with each other, forming the collectivity of the traditional Chinese culture.

The basic spirit of the traditional Chinese culture covers four main aspects, that is, people-oriented value system of the humanism; self-improvement, open-minded optimism of the national psychology; perceiving the concept of the object from an overall intuitive way of thinking; as well as the aesthetic ideal of harmony between man and nature. In respect of the value system, the Chinese culture represents the spirit of practical rationality emphasizing the people-oriented aspect.

The traditional way of thinking has two characteristics, the direct view of intuitive understanding and the symbolic view of perceiving the concept of the object. Harmony between man and nature is the aesthetic ideal and the highest realm of the traditional Chinese culture.

The traditional vernacular architecture of China is the embodiment of the traditional Chinese culture in the architecture. Through its layout and shape, we can feel the profound impact of the traditional Chinese culture and thinking. Mirroring the philosophy, patriarchal and environment concepts and thinking of the traditional Chinese culture, the Chinese vernacular architecture fully reflects, from different levels, the brilliant and profound wisdom of the traditional Chinese culture.

The impact of the Chinese traditional culture in the Chinese vernacular architecture, first of all, is the patriarchal view, that is, the family rituals. The core of the family rituals is hierarchy which has a strict requirement in vernacular houses in terms of their layouts, shapes, room spaces, scales, furnishing and decoration, so as to maintain the feudal hierarchy emphasizing blood ties.

The concept of harmony between man and nature comes in the second place, which serves as the basis for the orientation, location, and site selection of an architecture, as well as the principle for the building construction in a grand environment promoted by Confucianism and Taoism of the Han people. It includes architectural layout, integrity of space structure, order and domestication. In the ancient times, the ideal of harmony between man and nature in the Chinese vernacular houses was achieved through *fengshui*, or geomancy, that is, the five-orientation and four-god environment mode. The five orientations are east, south, west, north and center; and the four gods in four directions are Black Dragon, White Tiger, Phoenix, and Xuanwu.

Then is the way of thinking embodied in the Chinese vernacular houses, which fully reflects the spirit of humanism in the traditional Chinese culture. Everything is people-oriented, prioritizing the owner and family of the house. The trinity of Heaven, Earth and Man has Man at the first place. Thus, the residential construction is based on man's production and living.

Those are the influence and features of the traditional Chinese culture in the Chinese vernacular houses.

New Words and Expressions

heritage /ˈherɪtɪdʒ/	*n.*	any attribute or immaterial possession that is inherited from ancestors 遗产
substantial /səbˈstænʃl/	*adj.*	fairly large 大量的
abundant /əˈbʌndənt/	*adj.*	present in great quantity 丰富的，充裕的
vicissitude /vɪˈsɪsɪtuːd/	*n.*	a variation in circumstances or fortune at different times in your life or in the development of something 变迁，兴衰
quintessence /kwɪnˈtesns/	*n.*	the purest and most concentrated essence of something 精髓
exquisite /ˈekskwɪzɪt/	*adj.*	delicately beautiful 精致的
disparity /dɪˈspærəti/	*n.*	inequality or difference in some respect 差异，不一致
aesthetic /esˈθetɪk/	*adj.*	relating to or dealing with the subject of aesthetics 美学的
rite /raɪt/	*n.*	any customary observance or practice 惯例，仪式
acme /ˈækmi/	*n.*	the highest point (of something) 顶点
intuitive /ɪnˈtuːɪtɪv/	*adj.*	spontaneously derived from or prompted by a natural tendency 直觉的
harmony /ˈhɑːrməni/	*n.*	compatibility in opinion and action 和谐，和睦
embodiment /ɪmˈbɑːdɪmənt/	*n.*	a new personification of a familiar idea 体现
patriarchal /ˌpetrɪˈɑːrkl/	*adj.*	characteristic of an entity, family, church, etc., controlled by men 家长的，父权的
ritual /ˈrɪtʃuəl/	*n.*	any customary observance or practice 仪式，惯例
hierarchy /ˈhaɪərɑːrki/	*n.*	a series of ordered groupings of people or things within a system 等级制度
prioritize /praɪˈɔːrəˌtaɪz/	*v.*	assign a priority to 把事情按优先顺序排好
trinity /ˈtrɪnəti/	*n.*	a group of three 三位一体

Speaking and Writing

1. **Work in pairs to choose an interesting ancient house in your hometown or home village. One of you will be the tourist and the other the tour guide. The guide should outline the Chinese culture's impact on the house as well as introduce the general structure of the house. The tourist may have a few problems understanding the guide. Practise your dialogue and perform it in front of the class. These expression may help you.**

   Excuse me, I'm afraid I can't follow you.
   Please, can you speak more slowly?
   I beg your pardon?
   What did you mean by...?
   I didn't understand...
   I'm sorry but could you repeat that?

2. **Write down notes on the information you have gathered. You are going to write it in a local guide book. You want to encourage people to visit it so you should write in an exciting way.**

☞ **Model**

Why not visit Jiaotai Dian?

North of the Hall of Union (Jiaotai Dian), the Palace of Earthly Tranquility (Kunning Gong) was built for the chief consort of the emperor. During the Qing Dynasty, the Palace was remodeled into a Manchu-style house, which was dubbed "pocket house" (koudai ju): the house has its main door off center to the east rather than in the middle; wooden panel doors replace lattice doors; windows open from the bottom (swinging out on hinges fastened at the top) and are propped up by sticks. Inside the palace, along the north, the west, and the south walls are linked heated brick kang platforms. The Palace of Tranquility was at once the Shamanism sacrificial hall and the imperial bridal chamber. It still retains the original decor today.

3. **Summing up**

1) Write down what you have learned about Chinese traditional vernacular architecture.

____

____

2) From this unit you have also learned

- useful words: ____________
- phrasal verbs: ____________
- useful expressions: ____________

Unit 2

The Composition and Characteristics of the Chinese Traditional Vernacular Architecture

▶ **Pre-reading Activities**

1. Is there any traditional Chinese vernacular architecture in your hometown? Or have you seen any? What does it look like and what it is made of?
2. What do you have to consider when you want to build a house?
3. Is there any difference among Chinese northern vernacular architectures, southern ones, eastern and western ones? In which part of China you may find the vernacular architectures above?

Text A

The Composition of the Chinese Traditional Vernacular Architecture

The composition of the Chinese vernacular architecture involves four factors, that is, social, economic, natural and cultural factors.

Social factors, including productivity, social consciousness, ethnic differences, religious beliefs, customs and so on

China is a multi-ethnic country. Take the Han people as an example. In the feudal society which gave priority to patriarchal system and dominated the history for a long time, the domestic economy was based on small self-sufficient production and the clan was maintained through blood ties. The spiritual pillar of the social stability was the ethics doctrine of Confucianism, which promoted the moral values of seniority, brotherhood, man being superior to woman, and differentiated tasks for men and women, and advocated the family pattern of several generations living together under one roof as a symbol of prosperity for the family. As for the vernacular architecture, it must meet the needs of living and domestic production and prevent the family from external interference at the same time. Such a practice of self-isolation imposed, in particular, severe restrictions on women's activities which were confined to the inner yard of the residence. Another important element of the patriarchal system is ancestor and god worship, worshiping the clan ancestry and offering sacrifice to the gods in the region. Such patriarchal system and moral values have far-reaching impact on the vernacular houses in terms of the layout, room structure and the scale.

Hierarchy and family rituals of Confucianism, the core of the Chinese feudal system, gave birth to the courtyard-style vernacular houses of the Han people. The hierarchy of the feudal system was so stern that it had hard and fast rules on the architecture in terms of the scale, size, width, spatial room, as well as the roof form, and even the furnishing, decoration and color. For example, the houses for ordinary people should not have more than three rooms, with only plain black or white color; while the layout for large houses could be designed with several blocks, courtyards and roads, and even with study and garden. The layout of the vernacular houses exemplifies the influence of the social system on the architecture.

The economic factor is the material foundation in the formation of the vernacular houses

Materials are indispensable for the construction of a vernacular house in a certain configuration mode. Therefore, the quantity and quality of materials and their configuration determine the size, quality and grade of the architecture. The rich can decorate the doors, roofs and rooms in a very lavish way while the poor can only use mud walls and thin tiles to build their shelters.

However, the wisdom of the working people is unlimited. With local materials such as wood, bamboo, ash, stone, loess, they adapt the construction rules to meet their own needs based on the local natural conditions and their economic levels. Hence, the vernacular houses represent the most essential features of the architecture, having practical and reasonable functions, flexible designs, economic materials and simple appearances. In particular, for most houses, the builders are also the dwellers who design, construct and use these buildings. Therefore, the vernacular houses are people-oriented, cost-effective and practical, and most of all, reflect the ethnic features and the local characteristics of the region.

Natural factors include the climate, landform, terrain, materials and other natural substances and environmental factors

China borders vast seas on southeast and lies against steep mountains on northwest, extending about 5,500 kilometers from north to south and covering a width of about 5,200 kilometers from east to west. Within the territory, the high land in northwest lowers gradually southeastwards. In the west, there lies grandeur plateau; while in the east there is large plain. Mountains, hills, network of rivers and streams in the southeast are distributed in this wonderful land.

The climate in north and south China is quite different: in the north, winter is cold with frost and snow; in the south, it is very hot, humid and rainy in the summer and in some areas there is no snow all the year round. Each year in summer and autumn, the coastal areas in southeast are often harassed by typhoons which bring heavy rain and cause great damages to people, livestock and houses.

Moreover, due to the diversified geography and climate, the natural resources for construction materials are quite different. Loess is mostly found in the central plain and northwest region; hilly and mountainous areas boast prolific wood and stone; the southern region is rich in bamboo and brick; tile and natural gravel can be produced and collected in many places. In some coastal area, ash burned from shells can be found.

The vernacular house is constructed and completed in a given location and environment and under given geographical and climatic conditions. The dry and cold weather in the north and the hot and humid weather in the south lead to different approaches and methods in constructing vernacular architectures in different places. Taking geographical environment as an example, there are slope, flat land, river, creek and mountain. The houses built on the slope, flat land or beside the water present different views. The climatic factor has even greater impact on vernacular architecture in terms of the layout, design and interior space. That's why vernacular architecture in different places presents different forms and characteristics in various regions.

Cultural factors include folk situations, folk customs, production, ways of living, culture and aesthetic concepts

Cultural factors play a vital role in vernacular houses of the Han people, which is dominated by the Confucian ethics theory. The greatest impact on vernacular architecture is the thinking of worshiping the Heaven and Ancestry.

The first consideration for the design of a vernacular house is the ancestral temple or hall. It is stipulated in ancient rite that "for the house construction, priority should be given to the ancestral temple, second to the stall, then to rooms". In the chapter of Ancestral Hall Construction in *Family Ritual* composed by Zhu Xi of the Song Dynasty, which says "for the house construction, the ancestral temple, in three-room or more, should be situated at the east of the main bedroom". It shows that the ancient rite attaches great importance to and restrictions on ancestral temples.

The construction of a vernacular architecture which is used for both living and worshiping gives priority to worshiping. In the design, the ancestral temple where the family's ancestry and the Heaven and Earth are worshiped is placed in the middle of the last block of the entire residence. It is named as the rear-hall, also known as the ancestral hall. There are restrictions for the width and depth of the rear hall and the height of ridges and brims, so are the position and height of an altar and incense burner, which should not be changed at random.

Another thinking that influences the vernacular architecture is *fengshui*.

Fengshui, or geomancy, originating from the theory of *Yin-Yang* and Five Elements, is a theory in the ancient China, comprising climate, geography and environment for the location and orientation of Yang-House and Yin-House, respectively. The Yang-House is the vernacular

house. For instance, in rural areas, the site of houses in general has a relatively fixed pattern, holding Yang and lying against Yin. That is, a village should face the running water in front and have a high mountain behind. A house should face the south; the terrain should fall forwards. Such a layout is reasonable by the analysis of the modern concepts. For example, the running water lies in front of the village, meeting people's needs for fresh water, transport and washing; the high mountain behind it can be a perfect screen to resist cold wind; building a house on a tilting terrain keeps the house dry and makes it easy for drainage, which is ideal for living and health.

Fengshui contains symbolic and evil-suppression thinking. For example, the horse-head gable is quite popular for the vernacular houses in the region south of the Yangtze River and southern part of Anhui Province. The name given as the top of the gable is built into step-style canopy, which is shaped into a horse head in the forefront roof, thus known as the horse-head stepped gable. According to the local people, where the gable is shaped into horse-head, it indicates that there has been a *Juren* (a successful candidate in the imperial examinations at the provincial level in the Ming and Qing Dynasties) in the family. For a military officer, the gable is shaped into the horse-head and named as horse-head stepped gable; for a civilian official, the gable is built into seal or square shape and named as seal gable. All these shapes are actually show-off ways in the architecture to splurge that the family there had bred a *Juren* or an imperial official. For ordinary people, they can only use double-sloping roofs.

Likewise, in Chaozhou, Guangdong Province, the gable top is shaped in accordance with the practice of Five Elements of metal, water, wood, fire and earth. In the actual survey, two kinds of gables are usually adopted in the vernacular architecture: one is of curve shape called water gable; another is of pyramid shape called gold gable. In accordance with the Five-Element Theory, water suppresses fire while gold gives birth to water. Thus, the purpose of water and metal gable is to check and prevent fire. Since architectures in the ancient times were made of wood which was vulnerable to fire, once a building caught fire, it was very difficult to put the fire out. Because of the limited scientific knowledge, no effective measures could be taken for fire fighting, so people resorted to this psychological relief by checking the evil and praying for prosperity and safety. It well proves that the concept of Heaven and Earth has a profound impact on the vernacular architecture.

New Words and Expressions

clan /klæn/	*n.*	a group which consists of families that are related to each other 宗族
doctrine /ˈdɑːktrɪn/	*n.*	a set of principles or beliefs, especially religious ones 信条，学说
seniority /ˌsiːniˈɔːrəti/	*n.*	the importance and power that some people have compared with others 资历，辈分
differentiate /ˌdɪfəˈrenʃieɪt/	*v.*	recognize or show the difference 区分
advocate /ˈædvəˌkeɪt/	*v.*	recommend publicly 提倡，拥护
interference /ˌɪntərˈfɪrəns/	*n.*	unwanted or unnecessary involvement in some-thing 干涉，干预
impose /ɪmˈpoʊz/	*v.*	use authority to force people to accept something 强加
restriction /rɪˈstrɪkʃn/	*n.*	something that limits or controls something else 限制
ritual /ˈrɪtʃuəl/	*n.*	a religious service or other ceremony which involves a series of actions performed in a fixed order 仪式
stern /stɜːrn/	*adj.*	severe, strict 严格的
indispensable /ˌɪndɪˈspensəbl/	*adj.*	absolutely essential 不可或缺的
configuration /kənˌfɪɡjəˈreɪʃn/	*n.*	arrangement of a group of things 格局，布局
lavish /ˈlævɪʃ/	*adj.*	very elaborate and impressive 奢侈的，盛大的
loess /ˈloʊɪs/	*n.*	a light-coloured fine-grained accumulation of clay and silt particles that have been deposited by the wind 黄土
territory /ˈterətɔːri/	*n.*	land which is controlled by a particular country or ruler 领地
grandeur /ˈɡrændjər/	*n.*	impressive because of something's size, beauty, or power 宏伟，壮丽
plateau /plæˈtoʊ/	*n.*	a large area of high and fairly flat land 高原

harass /həˈræs/ *v.* trouble or annoy 骚扰，烦扰
boast /boʊst/ *v.* to have or possess 有，拥有
gravel /ˈgrævl/ *n.* consisting of very small stone 砾石
slope /sloʊp/ *n.* the side of a mountain, hill, or valley 山坡
dominate /ˈdɑːmɪneɪt/ *v.* to be the most powerful or important person or thing in something 占首要地位
stipulate /ˈstɪpjəleɪt/ *v.* say clearly that something must be done 规定
stall /stɔːl/ *n.* small enclosed area in a room which is used for a particular purpose, for example a shower（用于洗澡、沐浴的）小隔间
ridge /rɪdʒ/ *n.* 山脊，屋脊
altar /ˈɔːltər/ *n.* a holy table in a church or temple 祭坛
terrain /təˈreɪn/ *n.* 地形，地势
gable /ˈgeɪbl/ *n.* 山墙
canopy /ˈkænəpi/ *n.* a decorated cover 华盖
forefront /ˈfɔːrfrʌnt/ *adj.* the front or nearest the viewer 最前面的
suppress /səˈpres/ *v.* prevent something from continuing by using force or making it illegal 镇压

incense burner 香炉
have impact on 影响
attach importance to：consider something is important 重视
give priority to：treat something as more important than anything or anyone else 优先考虑，给……优先权
in accordance with：in the way that the rule or system says that it should be done 依据
resort to：have to do that action in order to achieve something 诉诸，求助于

Exercises

Task 1 Discovering the Main Ideas

1. Answer the following questions with the information contained in Text A.

1) What impact do patriarchal system and moral values have on Chinese vernacular houses?
2) How do hierarchy and family rituals of Confucianism influence the courtyard-style vernacular houses of the Han people?

3) What are the factors that make vernacular architecture in different places present different forms and characteristics in various regions?
4) What are the scientific ground behind "holding Yang and lying against Yin" pattern?
5) Why did ancient Chinese adopt water gable and metal gable in the vernacular architecture?

Task 2 Reading Between Lines

2. Fill in the blanks with the given words. You may not use any of the words in the bank more than once. Change the form of the given words if necessary.

involve	ethnic	dominate	doctrine	impose
advocate	indispensable	restriction	resist	accordance

1) Further work will ______ more detailed mapping, pitting and trenching.
2) Chocolates make me fat but I can't ______ them.
3) With globalization developing, no country can ______ the whole world.
4) He linked the ______ of natural law with the theory of popular sovereignty.
5) Parents should not ______ their own tastes on their children.
6) We ______ solving international dispute by negotiation, instead of appealing to arms.
7) We should make decisions in ______ with specific conditions.
8) China is a unified multinational country with 55 ______ minorities.
9) Privately-run schools are ______ parts of the education system.
10) The park is open to the public without ______.

Task 3 Challenge Yourself

3. Translate the following paragraph into Chinese.

Fengshui, or geomancy, originating from the theory of Yin-Yang and Five Elements, is a theory in the ancient China, comprising climate, geography and environment for the location and orientation of Yang-House and Yin-House, respectively. The Yang-House is the vernacular house. For instance, in rural areas, the site of houses in general has a relatively fixed pattern, holding Yang and lying against Yin. That is, a village should face the running water in front and have a high mountain behind. A house should face the south; the terrain should fall forwards. Such a layout is reasonable by the analysis of the modern concepts. For example, the running water lies in front of the village, meeting people's needs for fresh water, transport and washing; the high mountain behind it can be a perfect screen to resist cold wind; building a house on a tilting terrain keeps the house dry and makes it easy for drainage, which is ideal for living and health.

____

____

____

____

4. Suppose you are an expert on Chinese vernacular architecture, introduce the cultures embodied in the following pictures.

____

____

____

____

____

____

____

____

____

____

____

____

____

____

____

Text B

The Characteristics of the Chinese Traditional Vernacular Architecture

The characteristics of the traditional Chinese vernacular architecture mainly refer to the features most representing the people and the region, especially reflecting the production and ways of life, customs, beliefs, aesthetic concepts of various ethnic groups. Specifically, the characteristics of the vernacular architecture are reflected in the following three aspects.

Layout, mainly manifested in the rich planar and varied spatial combination

Take the vernacular houses of the Han people for example. Whether a large house with multi-courtyard cluster or a small house with three-in-one or four-in-one courtyard, the basic layout is the same, featuring fore-hall-rear-room, symmetrical central axis, two-room and main-hall and marked difference between the main and sub-rooms. The house is set in precise arrangement, featuring closed high wall outside and layer upon layer courtyard inside, developing in a vertical or horizontal direction, forming a flexible and compact layout being closed to outside and spacious inside. Dai people living in Yunnan, however, is an example for other ethnic groups. Every vernacular house in Dai villages is a single-family bamboo building because in the feudal territorial economic system where the monogamous family is prevailed, each household must pay tax and elder children must live independently after getting married. The inner is spacious and separated simply by wooden partitions, containing only sitting rooms or living rooms which also serves as the kitchen. Such arrangement mainly conforms to the way of life for small families. Although the Dai people believe in Buddhism, instead of setting up shrines for Household Gods in their houses, they build Myanmar Temple in the village for worship. In some bamboo houses there are shrines for Household Gods but such practice is indeed influenced by the Han people.

The outward appearance of the vernacular houses, mainly manifested in plain style, overall harmony, beautiful environment, as well as distinctive ethnic characteristics

Most of the Chinese vernacular houses are located in beautiful natural landscape in order to achieve comfortable living conditions. Architectures in rural areas are deployed along the natural terrain, either gathering together or distributing in a casual manner, either setting high or low. But in cities and towns, the architectures are adjacent to each other, constituting neighborhoods. For an individual building, it is designed to organize its space in a very flexible way on the basis of functional needs. Either in regular or random arrangement, the natural design reflects the simple and real features of the vernacular houses.

China has more than 50 ethnic groups which have experienced different historic events and enjoy diversified natural living conditions. The differences in their own production modes, customs, religious beliefs and aesthetic values result in great differences in construction forms in vernacular houses such as the thick-wall step-style flat-roof houses of the Tibetan people, the arcade-style flat-roof houses of the Uygur people, the stilt houses of the Dai people, etc.

In addition to the different customs, cultural faiths and aesthetic values of various ethnic groups, the hilly landscape, uneven terrain and great disparities in weather in China lead to diversified appearances of the vernacular houses in various ethnic groups which are representatives of the rich, splendid and distinctive ethnic characteristics.

The characteristics of vernacular houses' detail including furnishing, decoration, color, pattern, style, etc., mostly rooted in the ethnic customs, preferences, wishes and aesthetic concepts, are mainly manifested in the richness, decency, combination of art and practical use, and of profound cultural connotation

In detail, the most striking characteristics are of the gate, doors and windows, gable wall and some decorative elements, which easily draw the attention of the visitors as they are located in the most obvious and significant places of the architecture. Thus, over a long period, these parts have become an important content to represent the ethnic group and local characteristics.

In feudal society, the gate (or entrance door) was an important symbol to differentiate the rich and the poor, the high and the low. In the old society, all the families, whether rich or poor, would take every means to decorate and beautify their entrance doors to show their social status, making it an economic and cultural symbol of the vernacular houses, such as the entrance door and floral-pendant of the four-in-one courtyard houses in Beijing, the gate-building of the Bai people in Yunnan, and *tanglong* door (趟栊门) of a vernacular house in Guangzhou. Through the set-up and image of the gate, it is relatively easy to identify which ethnic group or region the vernacular house represents.

Windows are the most common and frequently used components in a building. Window, regardless of its size, style, color, or lattice pattern, craftwork, all mirror the taste and aesthetic psychology of the ordinary people. The form of a window is indicative of the ethnic group and region it belongs to, such as the ribbed and trapezoidal sash windows of the Tibetan, the strip and pointed arch windows of the Uygur, the removable windows in the courtyard houses of the Han people in north China, the wood-framed and colored-glass windows of the Cantonese, and the arch windows of the cave dwellings in the central plains region.

Colors, decoration, patterns and certain designs, as a result of the frequent use by the local residents, have become a unique means of artistic expression and characteristic symbol. For example, residents in the region south of the Yangtze River favor grey tile and white wall with tile being organized into the leaking-window pattern or various plants patterns; residents in south China are fond of black-brick wall decorated with pottery ridge; the Dai people love to use bamboos for decoration, in the patterns of elephants, nut trees, peacocks, the sun or the moon and other decorative designs.

Besides, the ordinary people of all ethnic groups often embody their own desires, beliefs, aesthetic concepts, the most desirable and favorite things, with realistic or symbolic approaches, in their residential decoration—patterns, colors, styles and other components, such as cranes, deer, bats, magpies, plums, bamboos, lilies, ganoderma, the 卍-pattern, and web design of the Han people; lotus flowers of the Bai people in Yunnan; elephants, peacocks, nut trees and other patterns of the Dai people. All these lead to the abundant and colorful ethnic characteristics of vernacular houses of various ethnic groups in different regions.

New Words and Expressions

connotation /ˌkɑːnəˈteɪʃn/ *n.* the ideas or qualities which a particular word or name makes you think of 内涵

indicative /ɪnˈdɪkətɪv/ *adj.* If one thing is indicative of another, it suggests what the other thing is likely to be 象征的

embody /ɪmˈbɑːdi/ *v.* to be a symbol or expression of that idea or quality 表现

Speaking and Writing

1. Work in pairs and outline the characteristics of the following gates. One is Beijing courtyard gate and the other is a traditional *tanglong* door in Guangzhou.

2. In the passage, the writer presents his statements first and goes on to support the statements with specific details. Then the writer briefly tells us the reasons. Now read the following passage and finish the tasks below.

Moreover, due to the diversified geography and climate, the natural resources for construction materials are quite different. Loess is mostly found in the central plain and northwest region; hilly and mountainous areas boast prolific wood and stone; the southern region is rich in bamboo and brick; tile and natural gravel can be produced and collected in many places. In some coastal area, ash burned from shells can be found.

1) In the above passage, the writer presents his statements first and goes on to support the statements with specific details. The structure employed is as follows:

General statement: Moreover, due to the diversified geography and climate, the natural resources for construction materials are quite different.

Specific detail: How is that natural resources for construction materials are quite different?

①Loess is mostly found in the central plain and northwest region.

②Hilly and mountainous areas boast prolific wood and stone.

③The southern region is rich in bamboo and brick.

④Tile and natural gravel can be produced and collected in many places.

⑤In some coastal area, ash burned from shells can be found.

2) There are several other paragraphs in the above two texts that follow the same writing pattern. Find them out.

General statement: ______

Specific detail: ______

______

______

3) Write a paragraph with a general statement supported by specific details and reasons on the following topic:

Climate's Impact on the Roof of Chinese Vernacular Architecture

______

______

______

______

______

______

______

3. Summing up

1) Write down what you have learned about Chinese traditional vernacular architecture.

______

______

______

______

2) From this unit you have also learned

- useful words: ______

______

- phrasal verbs: ______

______

- useful expressions: ______

______

Unit 3

The Characteristics of the Art of the Chinese Traditional Vernacular Architecture

▶ Pre-reading Activities

1. What may be the major art characteristics of Chinese traditional vernacular architecture?
2. How is art reflected in Chinese traditional vernacular architecture?
3. Is there any difference between the art of characteristics of southern vernacular architecture and northern one?

Text A

The Group Layouts and Image of Detached Houses

The characteristics of the art of the Chinese traditional vernacular architecture are mainly reflected in such four aspects as group layouts, detached architecture forms, spatial combinations and detail decorations.

The group layouts — harmonious and unified

The characteristics of the Chinese traditional vernacular architecture are as follows: several houses are grouped into one yard and several yards form one homestead; several homesteads constitute a lane, a street or a village, which in turn becomes a part of a town or a city. The image of the traditional vernacular architecture can not only be reflected by a single building but also by a courtyard, a lane, a street, a village or even a marketing town.

In the towns and cities of the southern plain areas of China, with a dense population, houses are mainly arranged in groups and adjacent to each other in good order, surrounded with lanes or streets on four sides with the characteristics of neat formation and closure. For the sake of facilitating production and transportation as well as economizing farmland, houses in the countryside are commonly built along rivers, roads or on slope land with a very good orientation and certain regularities.

Quite a lot of minority nationalities are living in the mountainous or hilly country. Their houses are generally built on altitudes in line, some are half way up a mountain and some are at the foot of a hill. Such houses have the characteristics of flexible arrangement, proper distribution and being harmonious with their surroundings. The enclosed houses built for defense in the Hakka's mountainous areas are huge in contour, representing the characteristics of the people's dignity, imposing boldness and straightforwardness. Some of them are detached; some are undetached; some are round and some are tetragonal in shape; still others are interlocked with two or more

round or tetragonal enclosed ones.

Vernacular houses in the river or lake areas are arranged along rivers or close to water with maximum use of the water surface. In the region south of the Yangtze River, houses face streets and with water on the back, with buildings accommodating roads and the trends of the rivers so as to create exquisite surroundings by displaying the distinguishing features of the region of rivers and lakes.

In the villages of the central part of the Guangdong Province where it is hot and rainy, the vernacular houses are in the comb-type layout because they are arranged in great density just like combs, both regular and in good order. A fishpond is commonly seen in front of a homestead with a bamboo grove behind the house. One or two banyan trees may be planted beside the rice fields. Both trees and the rice fields around them reflect the simplicity, naturalness of the houses and the rural scenery of the farmyards of the village.

In some mountainous areas of such provinces as Fujian, Jiangxi, Guangdong, Shanxi and Shaanxi, out of the needs of defense, the vernacular houses, commonly grouped into huge blockhouses or enclosed houses, which can be single-storey, two-storey or multi-storey houses, round or tetragonal in shape, externally closed but internally spacious, are huge in contour, displaying the characteristics of steadiness and roughness. Some rammed earth houses comprise several or even more blockhouses, leaving people a profound impression.

In the stockade villages inhabited by the Dong people in the regions of Guizhou and Guangxi in southwest China, an open ground is commonly marked off to build a tall drum-tower serving as the center of the public activities. The tower, the rain-proof bridge at the end of a village and the traditional vernacular houses of the Dong nationality together constitute a beautiful overall horizontal outline.

The image of detached houses — simple and authentic

The detached vernacular houses in the Han people-inhabited areas, especially in the rural areas, generally one-storey, three-bay, pitched-roof ones with white walls and grey

tiles are built. Two-storey houses wearing simple and authentic contours are mainly built in towns or cities that are densely populated. Some houses are casually decorated on the gate, the doors, the windows, the gable walls or the skew blocks for embellishment. Some detached vernacular houses in the mountainous areas or slope land wear a light and handy contour since they are laid out flexibly to fit the landform. Some two-storey or more-than-two-storey houses on the top of the slope land install a veranda or an overhanging eave on each storey. Some utilize form combination and projecting eaves with various forms. The overhanging houses in the provinces of Yunnan, Guizhou, Guangxi and Hunan are good cases in point.

The vernacular houses in the mountainous areas or slope land of Sichuan Province are built downhill to fit the landform or raised layer upon layer along the terrain and the roofing is also built in the same manner with the sense of rhythms.

The most conspicuous part of a detached house is the form of its roof, which is the manifestation of the unique combination of aesthetics and structure adaptive to the conditions of the local climate and geography with local building materials. Different house plane-surfaces formed by different natural conditions, with different materials in different structural forms and different living styles of different nationalities have generated various figures of roofs. Therefore, for a long time, the styles of roofs have become the main features of the traditional vernacular houses inhabited by different nationalities in different parts of China, such as the roofs of the "horse-head" stepped gable walls of the houses inhabited by the Han people in the region south of the Yangtze River, the roofs on the pot-ear-shaped gable wall houses in Guangdong, the flat roofs of the houses inhabited by the Tibetans, the dome-shaped top of the houses inhabited by the Mongolians, the dome roof of the houses inhabited by the Uygur people, the towering *xieshan*-style roof of the houses inhabited by the Dai people, and the big through-jointed water drip roofing of the houses in Sichuan Province.

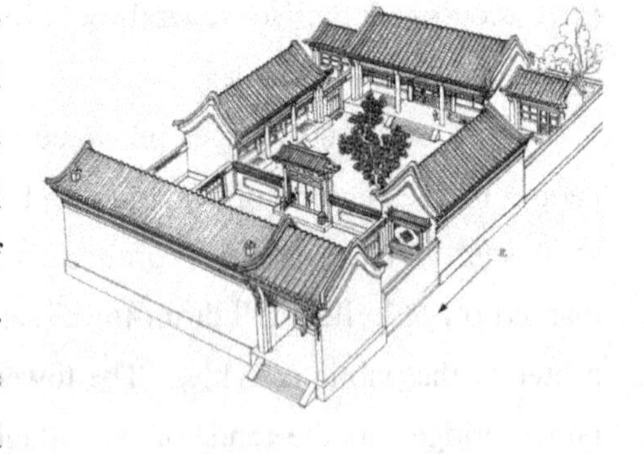

New Words and Expressions

detached /dɪˈtætʃt/ *adj.* not joined to any other house 独立的

facilitate /fəˈsɪlɪˌteɪt/	*v.*	to make easier 促进，使便利
Hakka /ˈhækə/	*n.*	a people of the Yellow River Plain that migrated into the hilly areas of southeastern China and Southeast Asia in the ancient time 客家
contour /ˈkɑːntʊr/	*n.*	outline of an object 轮廓，外形
tetragonal/ tɪˈtrægənəl /	*adj.*	square 方形的
rammed /ræmd/	*adj.*	forced or driven by heavy blows 夯实的
stockade /stɑːˈkeɪd/	*n.*	a line or wall of strong wooden posts 栅栏
authentic /ɔːˈθentɪk/	*adj.*	genuine 真正的，真实的
aesthetics /esˈθetɪks/	*n.*	a branch of philosophy concerned with the study of the idea of beauty 美学
gable wall		山墙
skew block		墀头

Exercises

Task 1 Discovering the Main Ideas

1. Answer the following questions with the information contained in Text A.

1) What are the characteristics of the art of the Chinese traditional vernacular architecture?

2) Why have the styles of roofs become the main features of the traditional vernacular houses inhabited by different nationalities in different parts of China?

3) What does the huge contour represent in the enclosed houses in the Hakka's mountainous areas?

4) How are the characteristics of harmony reflected in the Chinese vernacular architectures?

Task 2 Reading Between Lines

2. Fill in the blanks with the given words. You may not use any of the words in the bank more than once. Change the form of the given words if necessary.

detach	facilitate	authentic	characteristic	decoration
layout	harmonious	unify	constitute	conspicuous

1) This is an ______ news report. We can depend on it.

2) Windmills are a ______ feature of the Mallorcan landscape.

3) It's impossible to ______ oneself from reality.

4) The room was austere, nearly barren of furniture or ______.

5) They can build a more ______ society once inequality and exploitation are removed.

6) Volunteers ______ more than 95% of the center's work force.

7) The new airport will ______ the development of tourism.

8) She knows the ______ of the back streets like the back of her hand.

9) Its colouring makes it highly ______.

10) How can we ______ such scattered islands into a nation?

Task 3 Challenge Yourself

3. Look at the picture below and talk about the art characteristics of the vernacular houses presented.

4. Translate the following paragraph into Chinese.

Vernacular houses in the river or lake areas are arranged along rivers or close to water with maximum use of the water surface. In the region south of the Yangtze River, houses face streets and with water on the back, with buildings accommodating roads and the trends of the rivers so as to create exquisite surroundings by displaying the distinguishing features of the region of rivers and lakes.

____

____

____

Text B

The Space Composition and Detail Decorations

Space composition — flexible and graceful

The spaces of the vernacular houses consist of two parts — the interior and the exterior and the latter refers to the surroundings of the houses.

Sometimes, the artistic charm of the vernacular houses is expressed by combining the architectures with their surroundings instead of merely by the architectures themselves.

The hillside thatched cottages appear delicate and tranquil because they are located straight on tall hills; riverside houses look casual and elegant because they are situated beside murmuring streams and the houses on the slope land seem light and ingenious because they are arranged to suit the landform.

The art of the interior space of the houses inhabited by the Han people is reflected in two aspects: one is the space of the courtyard and the other is the space of both the hall and interior.

Space of the courtyard

Most of the vernacular houses of China center their spaces in the form of courtyards or halls. Take the four-in-one courtyard houses of north China for example. The principal room faces south and the chambers standing on the east and west wings are arranged symmetrically along the north-to-south axis. On the south, a gallery, a tracery and the secondary gate are set up. Outside the secondary gate is the front courtyard which is long and narrow from east to west and the entrance gate is set up at the southeast corner of the courtyard. Such kind of space compositions make a distinction between the primary and the secondary ones with clear and definite partition, not only satisfying the needs of serving to pay respect for seniority and to distinguish the inside from the outside, but also creating a quiet environment for residence.

The courtyards of the vernacular houses inhabited by the Han people in the south are rather small and so are generally called small courtyards. There are slightly larger courtyards commonly used as garden courts. Generally, no tree grows in small courtyards or garden courts because the trunks of big trees are thick, leaves dense, keeping out the sunlight and affecting ventilation. Sometimes one or two single thin trees or bamboos or potted plants may be planted but generally they are mainly for afforestation.

Medium-sized courtyards may have just one tree planted, complemented with a rockery or some greenery, linked together with the eaves gallery of the architecture to generate a quiet, easy and comfortable atmosphere.

Large courtyards can have rockery, ponds, flowering wood with a multi-storied pavilion, terraces or marble boats, linked with the residence to form a group of relatively integrate private gardens. With the natural landscape, a private garden can be divided into different scenic spots. By means of scenic focal point and borrowed view in the limited space, an artistic enjoyment can be achieved in residing, touring, walking and sightseeing.

The space inside the hall

First, doors and windows or partition boards are built in the place where the hall leads to the small courtyard directly. The design of a partition board is rich and varied, which is one of the components rich in artistic manifestation in a vernacular architecture.

In a large vernacular architecture, the panel of a door, a window or a partition board is usually shaped with trellis design, rattan-shaped lines and brocade-like lines or with the sculpture of figures or flowers and birds. The hatch of some grid-windows is added with a covering grid for keeping out the line of sight. Through large areas of designs, lines, and the comparison between the permeable shadows, the artistic effect of the decoration and finish is produced. In a small-sized vernacular architecture, jut window boards are commonly installed under windows, lintel eaves added to the top of windows, window bars sticking out if close to water, or bracket baluster on the storey to enhance the concavo-convex varieties and the void-and-solid contrast, by means of which the requirements for utilization are met and the artistic atmosphere is gained.

Second, the division of the hall and the indoor space is commonly created by adopting screens, shades or partition walls (or boards), which are characterized by carving techniques and fine designs on them.

Third, the hall's or veranda's beam mounts and their accessories such as girders, beam-ends, etc. are carved without affecting their structural property, to both enrich the spatial artistic effect and add embellishment.

Fourth, the horizontal inscribed board and couplet (written on scrolls and hung on the pillars of a hall) are indispensable for the hall. The old and well-known families, the respected and influential clans living in the Central Plains of the ancient China still maintained such family traditions after having moved to the south as hanging the name of the hall on an inscribed board on the wall or the lintel of the hall. A pair couplet may be hung on the golden pillars of the central bay of a hall. In addition, the horizontal inscribed boards and couplets may also be put up on the walls of a mansion, a small veranda, a storied building or a loft in a garden, which not only add the cultural tinges and poetic flavor to the garden but also enrich the artistic features of the garden's space.

Fifth, furniture and furnishing is the major means to enrich the indoor space as well as the essential tackling in a hall or housing. For example, tables, chairs, desks or tea tables are arranged in the hall for receiving guests; baldachins or shrines for ancestral tablets are set up for offering services to ancestors and the Heaven; pendent lamps or lanterns are installed for illumination at night. Wood beds, cabinets, wardrobes, tables and chairs are furnished in the bedrooms, which not only have practical values but also are decorated with exquisite carvings, full of artistic and national features.

Detail decorations — rich and varied

Detail treatment

Detail treatment occupies an important place in the artistic expression of the vernacular architecture, the purpose of which is mainly to enhance the artistic effect of the building. Detail treatment is usually adopted on practical locations, especially where people can directly see or touch with their hands. Locations of a vernacular architecture at display are usually the gate, the wall surface, the ground, the beam frames, the pillars, the staircases, the column plinths, banisters, steps, etc.

The gate of a vernacular architecture is always the symbol of the owner's social and economic status; therefore, people in different parts of the country manage the styles, materials, craft, decoration and colors of their gates very meticulously to highlight the family status.

The gate of a vernacular architecture in Beijing

is usually a detached entrance hall building. A two-leafed black gate is installed in the middle of the building, on which two iron rings are fixed. Without much decoration, it still looks serious and blocked, ghastly and bloodcurdling.

Some rich, scholar or literati-and-officialdom families south of the Yangtze River commonly use a *pailou* (牌楼) (an arch over a gateway) as their gate, on the lintel of which an inscription can be seen. Projecting eaves are made of grey bricks decorated with elaborate engraving to show their nobleness and refinement. Some age-old villages in the southern part of Anhui build memorial gateways as the entrance of the village. A typical example is the Tangyue Village in Shexian County, Anhui Province, in which one can see a memorial gateway before entering the village. The gates of the vernacular houses in some places are made in the form of recess (called the recesses gate), which effectively avoids being exposed to the sun and rain. The gate of a vernacular architecture in the central area of Guangdong Province is usually a black-varnished two-leafed one, outside which a fence gate made of some wood horizontal muntins is built, commonly called *tanglong*. When the gate is open, the fence gate is closed, with the purpose of both ventilation and security. Some families may install a four-leafed wood half-fence gate, which has a better effect on ventilation and keeping out the line of view. Such fence gates merge art and practicality into an organic whole with sturdy wood and beautiful designs.

The splay arch-gateway of the vernacular architecture of the Bai people is one of the most outstanding symbols of the architecture. The arch-gateway is splay outwardly and its wall is veneered with ceramic tiles. The well-proportioned roofing comprises the primary and secondary parts while the surface and the eaves of the wall are decorated gorgeously.

The gate of the multi-storied bamboo pavilion of the Dai people is set on the second floor. One can reach it via the covered way after climbing the wood ladder from the first floor. The gate is made of bamboo braid and some gates are woven with designs, simple but artistic.

Apart from the gate, many a Chinese vernacular architecture has a second gate leading to the main court, which is the symbol to differentiate the inside from the outside under the feudal system. Some rich and influential families are very particular about the second gate, such as the floral-pendant gate of the homestead in Beijing. In addition, the artistic door opening is mainly adopted in gardens. The technique of the building is light and free and the external form may be in the form of the moon, a circular, a bottle or a kettle.

Treatment to the wall surface is also one of the most important techniques of the artistic treatment to the detail of the vernacular architectures. The contrast effect is achieved by the quality and texture of the materials employed, such as the simple but elegant effect of the white plastering walls partitioned with maroon skeleton of wood posts in the areas south of the Yangtze River and East Sichuan. In the areas surrounding Quanzhou of East Fujian, stone blocks and red bricks are alternatively used to build walls, or stone frames are inlaid on red bricks; dark and light colored bricks are used to make different designs to achieve the contrast effect of the sense of quality and texture. In the area of Chaozhou, Guangdong Province, the artistic effect is achieved with varied profiles by making layers of lines on the drooping section of the squint of the wall. On the wall surface of Yang Amiao's (杨阿苗) courtyard in Quanzhou, Fujian, there is a brick carving and on the upper part of the wall there is another brick-carving summer eaves, which are graceful in design and consummate in carving craftsmanship.

In a vernacular architecture, especially in a garden with small rooms or verandas with windows, the ground of the courtyard and eaves galleries are commonly laid with bricks, tiles or stone. Some may be decorated with carving designs on the stone, and some with various designs pieced together with cobbles for embellishment.

The column base, which is also one of the parts to be decorated, has the functions of damp-proof, water-proof. It is often made in the form of a drum, a gourd, a cylinder, a bottle, a dipper, an octagon, etc. Some column bases have just a single layer and some double layers. The surface of some column bases is commonly carved with flowers, birds or animals while that of some others is carved with geometric figures. The craft and themes are characterized with obviously local colors.

须弥座式柱础　多角盆形柱础　如意纹饰柱础
铺底莲花柱础　仰覆莲花柱础　宝装莲花柱础

Banisters are generally made of wood or stone. The wood banisters are mainly used indoors, such as under the veranda, on the stairs, on the colonnade of the second floor, or in the parts suspended in midair such as the shaft, etc. to avoid being exposed to the sun and rain. The principle for the treatment is mainly practical combined with artistry. Banisters are practical and the artistic treatment to them gives first place to upholding simplicity with lines, decorated with adequate relief or low relief rather than go against the wood's structural behavior. The stone banisters are mainly used in outdoor sections close to water, such as a bridge over a ditch since the water is not wide. The stone banisters should be built short, small, solid and steady since the bridge may not be large. Close-water banisters in some gardens are made of stone battens, on which people can take a rest and look into the landscape on the opposite bank as an artistic enjoyment while taking a walk.

Lying on the ground, steps and platforms have the function of damp-proof. People will pay special attention to their steps when walking on them for fear of tumbling down since they are generally located in low places. Therefore, steps and platforms generally maintain their natural appearance without any treatment to the detail. If certain treatment is necessary, some lines may be simply carved on them.

Decoration and finishing

Decoration is a kind of added artistic treatment to the structural components, such as ridge ornament on the roofing, the decoration on the gate, the outer-eaves, the surface of the gable walls, and that on the interior beam frames, etc. It is practical, with the intent of embellishing the building without affecting the application and structure of it. It is also a symbol to show the owner's social status and wealth.

Decoration is also called small carpentry, mainly referring to the interior arrangement and furnishing, including doors, windows, partition boards and furniture, which have the values of practicality and appreciation.

Decoration is one of the important architectural means of artistic expression, the feature of which lies in making full use of the quality of the material and the techniques of the craft to carry out the artistic treatment so as to achieve the harmony and unity of the structural disposition and aesthetic perception.

The technological feature of decoration is to make full use of the tools such as knives, hammers, axes and saws to handle the material by directly making figures or artistic treatment with different methods according to different materials and thus to form different kinds of artistic expressions and styles of decoration.

Another feature of decoration is the feature of artistic conception. Its artistic expression is to make best use of the means of the traditional symbolism, the implied meaning and the wish of our country to combine the nation's philosophical theories, ethic thoughts and aesthetic awareness. Such kind of symbolism is commonly adopted in the decoration of the vernacular architecture by means of combining the form with sound or the implied meaning. By combining the form with sound, it means to take advantage of the homonym to obtain the symbolic effect through the images of some objects. For example, lotus (lotus is translated into *lian* (莲) in Chinese, the sound of which means in succession (连)) and fish (fish in Chinese is *yu* (鱼), the sound of which stands for enough (余)) stand for "having enough and to spare in succession". Bats, sika deer and peaches stand for felicity, good salary, and longevity. By combining the form with the implied meaning, the figure of a visual object is used to indicate the content of the implied meaning. For example, pine trees and cranes stand for longevity, peonies for gentility, lotuses for cleanness and plum blossom and bamboos for exemplary conduct and nobility of character. Some people may combine the form with sound and the implied meaning at the same time to express certain ideas. For example, adding the head of a *ruyi* (如意, an S-shaped ornamental object, usually made of jade, formerly a symbol of good luck) to the mouth of a bottle stands for safe and sound as one wishes. These designs and figures mainly reflect people's auspicious wishes as an incarnation of the cultural tradition and perception with national characteristics.

The Chinese vernacular architectures have plenty of techniques of decoration. Generally, decoration implements three principles: first, to combine practicality with artistry; second, to combine the structure with aesthetics; and finally, to comprehensively employ the special skills of other categories of national culture and art, such as painting, sculpture, calligraphy, horizontal inscribed board and couplet, etc. In this way, the national characteristics and the special artistic appeal of the decorative art is enhanced.

Besides, the techniques of the artistic expression in decoration and finishing have the following characteristics: the abundance and unity of the composition figures, the variety of themes and contents, not limited to one type, that is, historic figures, animals, plants, flowers and grass can all be employed in decoration. In the treatment of colors, the focus is put on elegance and simplicity while the major locations are highlighted more than others.

The arrangement for the locations to be decorated in a traditional vernacular architecture includes interior and exterior. The stress on exterior decoration is laid on the gate, the roof ridge, the gable walls and the screen wall while the interior decoration mainly focuses on the doors, the windows, the partition board, the beam frames and so on. The technological categories involve carving and engraving, including wood carving, brick carving, stone carving, ash sculpture, pottery sculpture, and clay sculpture. People living in the region following the line of the sea in east Guangdong are fond of using inlaid ceramic for decoration, which is effective on preventing the sea wind from hitting their houses.

Synthesizing the above-mentioned major contents and techniques of the artistic expressions of the vernacular architectures inhabited by the Han people in south China, we can sum up the general characteristics as follows:

1) The systematization of the layout, the richness of categories and the flexibility of the combination.

2) The Chinese vernacular architectures are prominent in adapting to local conditions, drawing on local resources and suiting certain requirements in accordance with the availability of materials under such natural conditions as the climate, geography, landforms, material, structure, etc.

This is because the architectures are built by the people, who are the designers, constructors and the users at the same time. Under the feudal system, the majority of the Chinese population was peasants and ordinary people in towns or cities, therefore their economic capabilities and requirements to the utilization of the architectures determined that the road of being diligent and thrifty in construction must be taken.

3) The contours are simple, groups harmonious, decoration and finishing varied and colorful.

Under the influence of being modest and not showing one's capabilities in the feudalist China, the contours of the traditional vernacular architectures were simple. It was hard to express its artistic behaviors in a detached building and so the artistic behaviors can only be incarnated in groups. The architecture was also influenced by the ideology of neutralization originated from the ancient times. For example, the central axis must be symmetric in the plane layout. In the treatment of the landforms, the front should be low while the back high, a building should face water and with hills at its back, surrounded on both right and left sides. The building must have a suitable contrast of sizes, good steadiness and balance in modeling the form of a building. All these indicate the characteristics of harmony. Apart from the symbols such as the scales, contours and bays, the hierarchical system of architectures was expressed to a large extent through decoration, finishing and details as the economic and social status of the users varied; the cultural qualities and the places people lived in were different. The superb and proficient skills of the craftsmen made it possible to express the perception of decoration and finishing. The decoration and finishing have great variety of categories, unique in conceptions, extensive in themes, varied in techniques, adding boundless brilliance and features to the artistic expressions of the Chinese vernacular architectures.

New Words and Expressions

thatched /θætʃt/	*adj.*	a thatched house or a house with a thatched roof has a roof made of straw or reeds 用茅草盖的
tranquil /ˈtræŋkwɪl/	*adj.*	quiet 安静的，宁静的
ingenious /ɪnˈdʒiːniəs/	*adj.*	showing inventiveness and skill 精巧的，设计独特的
ventilation /ˌventɪˈleɪʃn/	*n.*	allowing the air to enter and move 空气流通，通风设备
afforestation /əˌfɔːrəˈsteɪʃn/	*n.*	the process of planting areas of land with trees 植树造林
trellis /ˈtrelɪs/	*n.*	a light frame made of long narrow pieces of wood that cross each other 格子结构
brocade /broʊˈkeɪd/	*n.*	a thick, expensive material, often made of silk, with a raised pattern on it 织锦
concavo-convex	*adj.*	concavo on one side and convex on the other side 凹凸的

girder /ˈɡɜːrdər/	*n.*	a long, thick piece of steel or iron that is used in the framework of building and bridges 大梁
lintel /ˈlɪntl/	*n.*	a piece of stone or wood over a door 门楣
shrine /ʃraɪn/	*n.*	a place of worship 圣祠，圣龛
plinth /plɪnθ/	*n.*	a rectangular block of stone on which a statue or pillar stands 基座，底座
banister /ˈbænɪstər/	*n.*	a rail supported by posts and fixed along the side of a staircase 栏杆，扶手
ghastly /ˈɡæstli/	*adj.*	unpleasant or frightening 可怕的
bloodcurdling /ˈblʌdˌkɜːrdlɪŋ/	*adj.*	frightening and horrible 恐怖的
recess /ˈriːses/	*n.*	a part of wall that is set further back than the rest of the wall 壁龛
muntin /ˈmʌntən/	*n.*	window screening 窗格条
plastering /ˈplæstərɪŋ/	*n.*	paste made of sand, lime, and water which goes hard when it dry 灰泥
maroon /məˈruːn/	*adj.*	dark reddish-purple 褐红色的
drooping /druːpɪŋ/	*adj.*	hanging downwards 下垂的
squint /skwɪnt/	*n.*	looking in different directions 偏斜，斜视
consummate /kɑːnˈsʌmət/	*adj.*	perfect and complete in every aspect 完美的
colonnade /ˌkɑːləˈneɪd/	*n.*	a row of stone columns 石柱廊
tumble /tʌmbl/	*v.*	fall over with a rolling 滚落
carpentry /ˈkɑːrpəntri/	*n.*	the work of carpenter 木工手艺
felicity /fɪˈlɪsəti/	*n.*	great happiness and pleasure 幸福，喜悦

Speaking and Writing

1. Please match each letter that shows where each vernacular house is located with the the house in each picture.

A. Quanzhou area B. Chaozhou area C. Jiangnan and Sichuan East area

( ) ( ) ( )

2. Work in pairs and outline the artistic expression of the following decorations.

3. The texts include some paragraphs in which the writer starts with a general statement and then presents a number of supporting examples. Look at one of the paragraphs.

Another feature of decoration is the feature of artistic conception. Its artistic expression is to make best use of the means of the traditional symbolism, the implied meaning and the wish of our country to combine the nation's philosophical theories, ethic thoughts and aesthetic awareness. Such kind of symbolism is commonly adopted in the decoration of the vernacular architecture by means of combining the form with sound or the implied meaning. By combining the form with sound, it means to take advantage of the homonym to obtain the symbolic effect through the images of some objects. For example, lotus (lotus is translated into lian in Chinese, the sound of which means in succession) and fish (fish in Chinese is yu, the sound of which stands for "enough") stand for "having enough and to spare in succession". Bats, sika deer and peaches stand for felicity, good salary, and longevity. By combining the form with the implied meaning, the figure of a visual object is used to indicate the content of the implied meaning. For example, pine trees and cranes stand for longevity, peonies for gentility, lotuses for cleanness and plum blossom and bamboos for exemplary conduct and nobility of character. Some people may combine the form with sound and the implied meaning at the same time to express certain ideas. For example, adding the head of a ruyi (an S-shaped ornamental object, usually made of jade, formerly a symbol of good luck) to the mouth of a bottle stands for safe and sound as one wishes. These designs and figures mainly reflect people's auspicious wishes as an incarnation of the cultural tradition and perception with national characteristics.

General statement: Its artistic expression is to make best use of the means of the traditional symbolism, the implied meaning and the wish of our country to combine the nation's philosophical theories, ethic thoughts and aesthetic awareness. Such kind of symbolism is commonly adopted in the decoration of the vernacular architecture by means of combining the form with sound or the implied meaning.

Example 1: By combining the form with sound, it means to take advantage of the homonym to obtain the symbolic effect through the images of some objects. For example, lotus (lotus is translated into *lian* in Chinese, the sound of which means in succession) and fish (fish in Chinese is *yu*, the sound of which stands for "enough") stand for "having enough and to spare in succession". Bats, sika deer and peaches stand for felicity, good salary, and longevity.

Example 2: By combining the form with the implied meaning, the figure of a visual object is used to indicate the content of the implied meaning. For example, pine trees and cranes stand for longevity, peonies for gentility, lotuses for cleanness and plum blossom and bamboos for exemplary conduct and nobility of character.

Example 3: Some people may combine the form with sound and the implied meaning at the same time to express certain ideas. For example, adding the head of a *ruyi* (an S-shaped ornamental object, usually made of jade, formerly a symbol of good luck) to the mouth of a bottle stands for safe and sound as one wishes.

There are several other paragraphs in Text A and B that follow the above writing pattern. Find them out.

General statement: ______________________________

Specific examples: ______________________________

____

____

Write a paragraph with a general statement supported by examples on the Chinese vernacular architectures are prominent in adapting to local conditions.

4. Summing up

1) Write down what you have learned about the artistic characteristics of Chinese traditional vernacular architecture.

____

____

____

____

2) From this unit you have also learned

- useful words: ____
____
- phrasal verbs: ____
____
- useful expressions: ____
____

Unit 4

Tianjing Courtyard Houses (I)

▶ Pre-reading Activities

1. What do you think are the major differences between the climate and landscape of North China and South China? What impact do they have on the vernacular houses in the South and the North? You may use the following words for reference: rainfall, heat, network of rivers and streams, small stone bridges, large roof, heavy eave, corridor.
2. Look at the following pictures, one of which is a Beijing courtyard house and the other a tianjing (天井) courtyard in the South, and decide which one is the picture of a tianjing courtyard house. Then skim the text to see if you are right.

Text A

Vernacular Houses in South China

The layout of the Han people's houses in South China is basically the same as that of the courtyard houses in North China. However, since the southern region is damp and hot with plenty of rainfall, hilly land and rivers, dense population and tight land resources, a house site is generally called tianjing courtyard, which takes up little land, and its layout is compact, with buildings adjacent with one another.

The vernacular houses in South China have the following features:

First, the courtyards are small. Tianjing courtyard houses can offer a proper exposure to sunshine for the residences there who require the availability of sunshine but at the same time don't want to be exposed to it too much, considering the scorching heat in the South. In addition, tianjing courtyard, as an indispensable element of a homestead, is the place for ventilation, change of air, natural light and drainage, and serves as the convergence for the transportation within it.

Second, there are many lanes and alleys. The main passages connecting the clusters of the vernacular architectures are called lanes or alleys. There are several kinds of lanes and alleys: those in the open are called lanes, and some of such lanes may be called alleyways while those with capping are called verandas. Those in which you can catch sight of the sky and take shelter from the rain and the heat of the sun are called eaves galleries or open corridors while those from which you can not see the sky are called inner corridors. There are not only lanes and alleys that function as the connection among the buildings within a homestead, but also lanes and alleys that are called fire lanes, or passageways to connect homesteads with the function of fire prevention and ventilation.

Third, open halls and concealed chambers are the basic elements in the layout of the traditional vernacular architecture. As the hall in a home is the most important and indispensable public place as the core and representation of a kin culture, it is generally located in the central part, facing the small courtyard, and the partition board in front of the hall adopts the movable and open form. When something important occurs in the family, the partition board will be removed so that the hall and the small courtyard can be combined as one bigger space to accommodate more clansmen.

Grouped flexibly with plentiful types, the vernacular architecture can be formed longitudinally or horizontally or in a crisscross pattern. Tianjing courtyards are also afforested, the exterior wall is enclosed while the interior of the homestead is spacious so that the spaces of the exterior and the interior of the house can be tightly connected with each other.

The vernacular architecture in South China generally has very good orientations so as to obtain a better condition of facing the wind. At the same time, a very good ventilation system is formed through an open hall, a gallery or corridor and a tianjing courtyard. Practices have proved that this is the main reason for the small-courtyard-type houses to continue in the South.

The artistic manifestation of the tianjing houses mainly depends on the art of architectural complex, the treatment to the space, the impression about the quality of the material and the furnishing of the furniture as well as the decoration and finish.

New Words and Expressions

drainage /ˈdreɪnɪdʒ/	*n.*	the system or process by which water or other liquids are drained from a place 排水系统
convergence /kənˈvɜːrdʒəns/	*n.*	the act or process of converging; the tendency to meet in one point 汇合处
afforest /əˈfɒrɪst/	*v.*	growing trees 绿化
spacious /ˈspeɪʃəs/	*adj.*	large in size or area 宽敞的
partition board		隔扇

Exercises

Task 1 Discovering the Main Ideas

1. Answer the following questions with the information contained in Text A.

1) What are the three distinctive features of South China courtyard houses?
2) What are the major differences between South China courtyard houses and North China courtyard houses?
3) What are the major functions of lanes in the courtyard houses in the South?
4) What is special about the partition board in South China courtyard houses?
5) What is the main reason for small-courtyard houses to long endure in the South?

Task 2 Language Focus

2. Fill in the blanks with the given words. You may not use any of the words in the bank more than once. Change the form of the given words if necessary.

damp	layout	compact	adjacent	exposure
scorching	indispensable	drainage	accommodate	conceal

1) A hotel ______ to the new China Central Television (CCTV) headquarters in Beijing caught fire Monday night, witnesses said.
2) There is a direct correlation between ______ to sun and skin cancer.
3) This room is not big enough to ______ so many people.
4) Damp proofing control systems help fight against problems such as woodworm & wood rot from ______.
5) The kitchen was ______ but well equipped.
6) Peasants are busy weeding under the ______ sun.
7) Discipline and freedom form a unity of opposites; both are ______.
8) Decorative ______ of the room is so close to nature, all use of modern hand-painting.
9) He could not ______ his annoyance at being interrupted.
10) The rainy season is coming. It's a matter of great urgency to repair the ______ and irrigation equipment.

Task 3 Challenge yourself

3. Look at the pictures below and talk about the benefits, functions and usages of small courtyards and partition boards.

4. Translate the following paragraph into Chinese.

The layout of the Han people's houses in South China is basically the same as that of the courtyard houses in North China. However, since the southern region is damp and hot with plenty of rainfall, hilly land and rivers, dense population and tight land resources, a house site is generally called tianjing courtyard, which takes up little land, and its layout is compact, with buildings adjacent with one another.

Text B

Dongshan Engraved Building of Suzhou, Jiangsu Province

Dongshan Engraved Building of Suzhou was built in 1924. It is called the engraved building because its beams and purlins, doors and windows and the gate of the main building are engraved beautifully and exquisitely. The architecture is really a piece of cream in the Chinese engraving art since it merges such crafts as brick carving, wood carving, color decoration and clay sculpture into an organic whole as a marvelous creation excelling nature, exquisite without compare.

The engraved building faces east, and its plane is of multilateral four-in-one courtyard, with two rows of the width of five rooms, in which there is a gateway, the main building and the garden. In front of the entrance gate, there is a zigzag shadow wall on which an exquisite brick carving with two Chinese characters "Hong Xi (meaning grand auspiciousness)" is inlayed.

The frontage of the arch-gate is a single slope roof made of tilts and a granite gate inlayed with the paste-up of levigated blue bricks. The upper, middle and lower frames of the arch-gate are all engraved with brick reliefs whose brick-carving in relief, circular engraving and fretwork are harmonious and fused mutually. The main hall of

the front building has the most carvings, whose wood carving decorations are rich with exquisite crafts.

The whole building fully reflects the essence of the folk traditional carving art south of the Yangtze River, which fuses architecture, engraving, painting and calligraphy, literature, craft and gardening into an organic whole, and so it is praised as "The First Building South of the Yangtze River".

New Words and Expressions

purlin /ˈpɜːlɪn/	*n.*	a longitudinal member in a roof frame 檩条
merge /mɜːrdʒ/	*v.*	combine or come together to make one whole thing 融入
paste-up/ˈpeɪstˌʌp/	*n.*	a composition of flat objects pasted on a board or other backing 拼贴
levigate /ˈlevəˌgeɪt/	*v.*	make something smooth or polished 磨光
fretwork /ˈfretwɜːrk/	*n.*	wood or metal that has been decorated by cutting bits of it out to make a pattern 透雕细工
zigzag shadow wall		"之"字形影壁
granite gate		花岗岩大门

Speaking and Writing

1. Work in pairs.

Suppose one is a tourist and the other is a tour guide. The guide is supposed to introduce Dongshan Engraved Building. The following opening pattern is for your reference.

Good morning, everyone. Welcome to ______. We'll be arriving in ______ within an hour. In the ______, it is my pleasure to show you ______.

2. Next write down notes on the information you have gathered from Section B. You are going to write it in a local guide book. You want to encourage people to visit it so you should write in an exciting way.

____

____

____

____

____

3. Summing up.

1) Write down what you have learned about Chinese traditional vernacular architecture.

____

____

____

____

2) From this unit you have also learned

- useful words: ______________________________

____

- phrasal verbs: ______________________________

____

- useful expressions: ______________________________

____

Unit 5

Tianjing Courtyard Houses (Ⅱ)

▶ Pre-reading Activities

1. Have you been to Zhejiang Province? What are the major features of its landscape?
2. Discuss in small groups what impression you have on Zhejiang vernacular houses. You may use the following words for reference: ingenious layout, exquisite structure, courtyards linked by corridors, large roof and beams, exquisite carvings.

Text A

Some Tianjing Courtyard Houses in Zhejiang Province

The Lu's Former Residence in Dongyang City, Zhejiang Province

The Lu's Former Residence (卢宅) is located outside the east gate of Wuning Town (吴宁镇) of Dongyang City, which is diversified by hills and mountains, and a crisscross network of rivers and streams. Facing hills and mountains and surrounded with water, the residence has a very elegant environment.

The Lu's Former Residence was set up during Emperor Yongle's (永乐) reign of the Ming Dynasty (1403 – 1424) and the existing houses were built in succession during the Ming and the Qing Dynasties. The Lu's Former Residence is an architectural complex consisting of several north-to-south vertical axes, with several thousands of houses on an area of 150,000 square meters. The main architecture is the Suyongtang Complex (肃雍堂建筑群) at the north end of the street.

Entering the arch gateway of the Former Residence of Lu (a high official in Feudal China, 卢大夫), you can see a double-leaf black-lacquered gate and double-layer brick overhanging eaves of the houses against a grayish wall, with a very simple and plane appearance.

The Suyongtang Hall is the public hall for the big clan of the Lu's, built in the Ming Dynasty, with a very prominent scale and social status. It is as wide as three bays, with high eaves and a high ridge. The main hall is connected with the back hall by a hallway, forming a 工-shape plane. The Suyongtang Complex consists of nine yards as the most completely preserved axis in the Lu's Former Residence. The craft of such wood-carving decorations as the beam mounts, the pillar columns, the cornices, etc. is exquisite with rich themes and harmonious designs, reflecting the strong characteristics of the local culture.

Zhuge Village, Lanxi City, Zhejiang Province

Zhuge Village (诸葛村) is situated in Lanxi City, Zhejiang Province and once named Gaolong (高隆) in ancient times. After the descendants of Zhuge Liang (a statesman and strategist during the period of the Three Kingdoms, 诸葛亮) settled there, the village was planned and constructed according to the Nine-Palace-Eight-Diagram battle formation created by their ancestor Zhuge Liang.

Zhuge Village boasts itself for a very good natural environment with high hills behind and plains in front of it where people can make a living either by farming or by gathering firewood, by fishing or by hunting. Furthermore, it has a very convenient transportation so that it is appropriate for inhabitancy. The village is surrounded with 8 hills and its layout takes Zhong Pond (钟池) as the center and the houses are of the radial form. The 8 lanes spreading outward divide the whole village into 8 sections so as to form the pattern of interior Eight-Diagram battle formation. Zhong Pond in the center of the village, a big pond of Eight-Diagram battle formation, looks like a diagram of supreme pole with Yin and Yang, the half with water as Yin

and the other dry half as Yang. The water in the pond adds mystical coloration and vivid anima to the village.

More than 200 vernacular houses built during the Ming and Qing Dynasties exist in Zhuge Village, whose layouts are ingenious and structures exquisite. The Ancestral Temple of the Prime Minister (in ancient China) was built during the reign of Emperor Wanli (万历) of the Ming Dynasty (1573 – 1620). Four logs with the diameter of 50 centimeters (a pine, a cypress, a tung and a Chinese toon) were selected for building the Middle Hall, with the implied meaning of Song Bai Tong Chun (meaning "may you live long and remain strong like the evergreen pine and cypress", 松柏同春). The wing corners are all raised with exquisite carvings, solemn looks and imposing momentum.

Lishui Street, Yantou, Yongjia, Zhejiang Province

Lishui Street of Yantou, located on the west bank of the middle reaches of Nanxi River, was started to be built in the early Tang Dynasty. Tahe Temple (塔河庙), composed of the temple itself, the river, an opera stage, an official-reception pavilion (接官亭), age-old trees etc., is a large-scale outdoor activity center for the local villagers. It deserves the name of a landscape garden.

On the reservoir embankment in East Park, there built an ancient commercial street — Lishui Street. It is more than 300 meters long, with 90 multi-faceted one-story or two-story shops, lined up across the river. Near the water is the *meirenkao* (a chair wins its name for being seated by beauties and leaned along their waist, 美人靠). In the history, the place was the only passage for salt merchants, who could have a rest or do business here. It is also a very unique street with a pleasant landscape.

At the south end of the street is the South Gate and there is a pavilion on the terrace nearby. Fifty meters away from the pavilion is a sound and delicate flower pavilion (also known as the official-reception pavilion), four-sided and two-story, wooden wall on the first floor and out-stretching eaves on the second floor. With transparent flower windows and heavy eaves, the pavilion features simplicity and solemn elegance, matching well with Lishui Street.

New Words and Expressions

diversify /daɪˈvɜːrsɪfaɪ/	*v.*	to make something diverse 使多样性
crisscross /ˈkrɪsˈkrɒs/	*n.*	a pattern or design consisting of lines crossing each other 纵横交错
reign /reɪn/	*n.*	the period during which a monarch is sovereign 统治期间
succession /səkˈseʃn/	*n.*	existing or happening one after another 接连发生
anima /ˈænɪmə/	*n.*	inner self 灵魂
momentum /moʊˈmentəm/	*n.*	a longitudinal member in a roof frame
solemn /ˈsɑːləm/	*adj.*	very serious rather than cheerful or humorous 严肃的，庄严的

Nine-Palace-Eight-Diagram battle formation 九宫八卦阵式

Exercises

Task 1 Discovering the Main Ideas

1. Answer the following questions with the information contained in Text A.

1) When and for what purpose was Suyongtang Hall of the Former Residence of Lu built? What was its most important feature in layout?
2) What is special about Zhuge Village's layout and Zhong Pond?
3) Where is Lishui Street located and what was it built for?
4) What is *meirenkao*?

2. Fill in the blanks with the given words. You may not use any of the words in the bank more than once. Change the form of the given words if necessary.

diversify	locate	surround	reign
consist	reflect	exquisite	solemn

1) The three white marble sculptures have been made with ______ craftsmanship.
2) The small churchyard is ______ by a rusted wrought-iron fence.
3) Oxford University is ______ in the northeast of London, about one hour's drive from downtown.
4) Archaeologists have dated the fort to the ______ of Emperor Antoninus Pius.
5) Hills and woods ______ the landscape.
6) Her sad looks ______ the thought passing through her mind.
7) The development will ______ of 66 dwellings and a number of offices.
8) Together with the characters on the wall, the picture makes the temple ______.

Task 2 Read and Simulate

3. The following English sentences on the left show some different ways of clarifying locations. Read the English sentences carefully and translate the Chinese sentences on the right into English by simulating the structure of the English sentences.

Sample Sentences	Translation
1) The Lu's Former Residence is located outside the east gate of Wuning Town of Dongyang City, which is diversified by hills and mountains, and a crisscross network of rivers and streams.	位于楠溪江中游的岩头镇，因位于芙蓉山脚下而得名。
2) Zhuge Village is situated in Lanxi City, Zhejiang Province and once named Gaolong in ancient times.	南屏村位于黟县县城西南4公里处。
3) Facing hills and mountains and surrounded with water, the residence has a very elegant environment.	面对正厅有精细的砖雕门楼。

续上表

Sample Sentences	Translation
4) On the reservoir embankment in East Park, there built an ancient commercial street—Lishui Street.	各组建筑入口均有石匾门额。
5) At the south end of the street is the South Gate and there is a pavilion on the terrace nearby.	村北是古渡码头，曰“淮溪首济”。

4. The following English sentences show some different ways of clarifying the construction time of a building. Read the English sentences carefully and translate the Chinese sentences by simulating the structure of the English sentences.

Sample Sentences	Translation
1) The Lu's Former Residence was set up during Emperor Yongle's reign of the Ming Dynasty (1403-1424).	钟楼村位于从化市太平镇，为欧阳氏族人在清咸丰年间所建。
2) The existing houses were built in succession during the Ming and the Qing Dynasties.	池东北端的“水月堂”始建于宋徽宗宣和年间（1119—1125 年）。
3) The village was planned and constructed according to the Nine-Palace-Eight-Diagram battle formation created by their ancestor Zhuge Liang.	承志堂建于咸丰五年（1853 年），为三进两层的木结构建筑。

Task 3　Challenge Yourself

5. Translate the following paragraph into English.

资政大夫祠古建筑群位于广州市花都区新华镇三华村的西面，它建于清代同治二年至三年（1863—1864 年），至今已有 150 多年的历史。

Text B

Some Typical Tianjing Courtyard Houses in Other Southern Provinces

Nanping Village in Yixian County, Anhui Province

Nanping Village（南屏村）, located four kilometers southwest of Yixian County（黟县）, is named after the mountain it lies against in the southwest.

Boasting high walls and deep lanes, Nanping Village has more than 1,000 villagers and there are 36 wells, 72 lanes and more than 300 vernacular houses of the Ming and Qing Dynasties. Lanes in different lengths interlinked with each other and running in all directions are found everywhere its village, winning its name of "Labyrinth in the South Yangtze".

Even today, a considerable scale of temples, shrines and ancestral halls still stand in Nanping Village which is known as the architectural museum of Chinese ancient shrine. The Ancestral Hall of the Ye clan "Xuzhi Hall"（叶氏叙秩堂）, located in the center of Nanping Village and facing west, was built during the reign of Emperor Chenghua（成化）of the Ming Dynasty. A pair of huge stone drums carved out of the bluestone of Yixian County about a person's height stand on both sides of the gate, glossy and dignified. From the entrance hall, the incense case in the main hall can be beheld and the whole building is spacious, bright and deep, creating a sudden and devastating silence and mystery.

The Banchun Garden（半春园）at the entrance of the village, built during the reign of Emperor Guangxu（光绪）of the Qing Dynasty, is a private school courtyard. There are planted a large number of rare flowers and trees, especially plum for the most, thus it is also named the Plum Garden.

Likeng Village in Wuyuan County, Jiangxi Province

In Jiangxi Province, "Keng" （坑） means a brook. All the villages in Wuyuan are built along the transparent and zigzag brooks. Likeng Village （李坑村） of Qiukou Town, situated in a narrow mountainous dock, was built in the Northern Song Dynasty. It lies in the arms of green mountains, enjoying a beautiful scenery. Most of the more than 260 households in the village inhabit along the river and a bluestone bridge is built in front of each house by the water, creating a vivid picture of a small bridge by the water, a flowing stream, and residence cottages.

Likeng vernacular houses belong to the Hui-school architecture. Apart from the typical features of the Hui-school architecture such as the powder wall, the eaves and cornices, the carvings of wood, stone and brick can also be called three marvelous techniques. The ancient buildings from the Ming and Qing Dynasties can be found anywhere in the village, which are constructed along the Cangzhang Mountain, covered with eye-pleasing walls and tiles. They are scattered irregularly. Within the village, streets, lanes and streams traverse and cross each other. There are crisscross roads paved with bluestone and dozens of stone, wood and brick bridges fly over the stream, creating a splendid picture-scroll of small bridge, flowing water and houses. It is a brilliant pearl in the ancient villages in Wuyuan.

Peitian Village in Liancheng County, Fujian Province

Peitian Village（培田村）is located at the Wujiafan（吴家坊）on the upper stream of the Heyuan River（河源溪）, Xuanhe Town（宣和乡）, Liancheng County, in the mountainous area of the west Fujian Province. With a history of over 800 years, it is a well-preserved ancient Hakka（客家）vernacular complex of the Ming and Qing Dynasties. Consisting of the private schools, the ancestral monuments, the ancestral halls, an old street of one-thousand-meter long, five lanes and two rivers running through the village, the village has an exact layout and reasonable arrangement, boasting a spectacular harmony.

Peitian ancient vernacular complex is represented by the "Residence of Minister"（大夫第）, the "Yanqing Hall"（衍庆堂）and the "Official Hall"（官厅）, which are famous Hakka architectures of "nine-hall-and-eighteen-patio" style. The "Residence of Minister", also known as the Jishu Hall（继述堂）, was built in 1829 and completed 11 years later. The Yanqing Hall is a building of the Ming Dynasty, whose structure is similar to that of the Residence of Minister. There is a lotus pond and the zigzag path in front of the building and solemn stone lions stand at the gate. The Official Hall is given such a name because it was used to receive passing-by officials. It has high walls and enclosed yard and presents delicate design and sophisticated workmanship. The flower halls on the left and the right sides are dedicated to entertaining its owner and his friends. The hall downstairs serves as the private school and the hall upstairs for storing books. Although there are a number of halls, rooms and tianjing, the vernacular house is arranged in an orderly manner in the layout. Halls are connected by paths but with independent doors, which makes it convenient both for the big family to live under one roof and small families to enjoy their own leisure and union.

The Vernacular Houses of Fenghuang County, Hunan Province

The historic town of Fenghuang County got such a name because the mountain on the southwest looks just like a phoenix (the bird is pronounced as "*fenghuang*" in Chinese). From the ancient times, the ancestors of the Tujia and the Miao ethnics began to live here and it has been about 3,000 years.

The historic town of Fenghuang County is situated by the Tuojiang River and the vivid water winds its way along the city wall through the town. Smoke is curling upward from the kitchens of the overhanging houses by the river, with a slim and graceful sight full of rhythms. The existing red sandstone wall of the town was built during Emperor Kangxi's reign of the Qing Dynasty. The town is situated at the foot of the mountain beside a stream and the houses in it are adjacent to each other regularly.

The layout of the houses in the town mainly follow the traditional courtyard pattern. For example, Shen Congwen's former residence is a typical tianjing courtyard one in south China, consisting of a front and a back yard, with through-jointed wood structure, and the horse-head gable wall decorated with the figure of the head of a huge legendary turtle. The architecture is small but untraditionally shaped, with ornamental engraving, having an antique, elegant and quiet flavor.

In the Chen's Historic Courtyard-shape Residence (陈氏老宅), typical in South China, there is a winding corridor around the small yard, on the right side of which there is a wood staircase leading upstairs. The carved doors and windows in the residence are elegant and exquisite and the interior decorative craft is fine and meticulous.

New Words and Expressions

labyrinth /ˈlæbərɪnθ/ *n.* a place made up of a complicated series of paths or passages 迷宫

traverse /ˌtrəˈvɜːrs/ *v.* to cross an area of land or water 横跨

sophisticated /səˈfɪstɪkeɪtɪd/ *adj.* more advanced and complexed than other 精密的

overhang /ˌoʊvərˈhæŋ/ *v.* stick out over or above something 悬挂

meticulous /məˈtɪkjələs/ *adj.* marked by extreme care in treatment of details 精细的

Speaking and Writing

1. Work in pairs. Suppose one is a tourist and the other a tour guide. The guide is supposed to introduce the location and history of Nanping, Likeng, Peitian, and Fenghuang County respectively by using the pattern learned in the exercise of Text A.

____

____

____

____

____

____

____

2. Summing up.

1) Write down what you have learned about Chinese traditional vernacular architecture.

____

____

____

____

2) From this unit you have also learned

- useful words: ____

____

- phrasal verbs: ____

____

- useful expressions: ____

____

Unit 6

Chinese Traditional Courtyard Houses

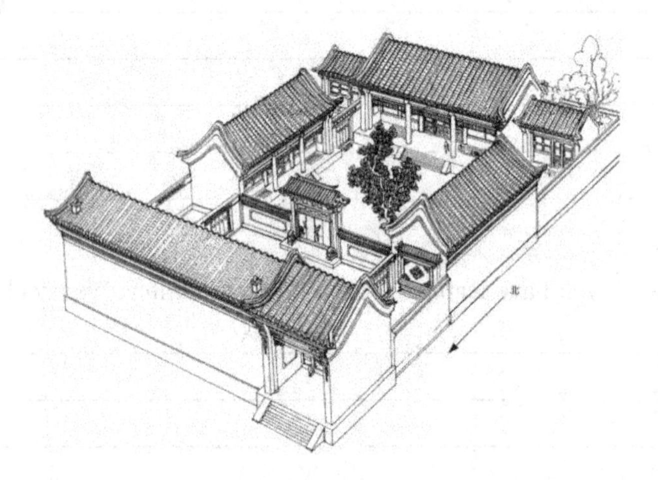

► **Pre-reading Activities**

1. What do you think are the major differences between a Beijing courtyard house and a tianjing courtyard in the South?
2. Look at the above picture. How do you feel about the layout of the yard? In what ways does it reflect our traditional culture?

Text A

Introduction to Chinese Traditional Courtyard Houses

The spirit of rites and ethics occupy a leading position in the Chinese traditional vernacular architecture of the Han nationality. Planes of multiple-courtyard houses are main patterns in the plane layout. Since China is characterized by a large territory and disparities in climates, the open-ground vernacular architectures can be classified into two main types: the northern type and the southern type. The former refers to the courtyard houses while the latter small courtyard houses because the southern areas have a large population but little land and it is humid and rainy there so that houses are adjacent to each other with small courtyards.

The courtyard house in north China consists of such detached buildings and components as the main block, wing-blocks, side blocks, the front block, the posterior block, the entrance gate, the floral-pendant gate, the U-shaped corridor, short corridors (in the corners), the covered corridor (leading from one court to another), the shadow wall, courtyard wall, the combination forms of which can have just one yard, two yards, three yards or more. Generally, a homestead is arranged longitudinally along the central axis. A large homestead may have a side yard along the central axis or two yards may stand side by side. Some may be arranged staggeringly according to the landform.

The plane layout of a four-in-one courtyard is that the central kernel courtyard is in the north-south orientation, one main block with two wing blocks added by a floral-pendant gate (垂花门) or a passing hall forming a pattern which is approximately a square. The main block, wing-blocks and the front block are distributed on the perimeter, not a single one is connected with anyone else. The central axis of the plane combination of the four-in-one courtyard is symmetric and the courtyard is spacious with abundant sunshine. The main block is the principal building, comprising the living room and bedrooms of the owner and his wife. The wing-blocks are located on both sides of the main block, generally serving as the bedrooms of the owner's children.

The front block is built on the frontage of the four-in-one courtyard, the front eaves of which face the interior courtyard and the back eaves facing a lane (or street). The entrance gate is located on the east of the front block, which has no window at all or just a high-light window. Therefore, it assumes a look full of the sense of closure.

The last block of the courtyard is the "posterior block", which is used for sundry duties such as the frontage, and the backdoor may be set up there.

The characteristics of the courtyard houses in north China are front-hall-rear-room, the central axis being symmetric, the plane being regular and tidy, the exterior being prudent and enclosed, reflecting a profound hierarchical perception.

New Words and Expressions

longitudinally /ˌlɑːndʒɪˈtuːdɪnəli/	*adv.*	extending in the direction of the length of a thing 纵向地
posterior /pɑːˈstɪriər/	*adj.*	situated at the back of something 后面的，背部的
staggeringly /ˈstæɡərɪŋli/	*adv.*	in a staggering way 令人吃惊地
kernel /ˈkɜːrnl/	*n.*	the central and most important part of something 中心，核心
sundry /ˈsʌndri/	*adj.*	different from each other 各式各样的

Exercises

Task 1 Discovering the Main Ideas

1. Answer the following questions with the information contained in Text A.

1) What spirit occupies a leading position in the Chinese traditional vernacular architecture of the Han nationality?

2) In a four-in-one courtyard, what orientation is its central kernel courtyard built in?

3) What does the main block comprise?

4) What is the function of the wing-blocks?

5) What are the characteristics of the courtyard houses in north China? And what do these characteristics reflect?

Task 2 Language Focus

2. Fill in the blanks with the given words. You may not use any of the words in the

bank more than once. Change the form of the given words if necessary.

occupy layout characterize classify axis homestead status corridor reflect

1) ______ Wall Street, a non-violent leaderless resistance movement fighting corporate greed and economic inequality, has picked up steam since its inception on September 17.
2) Perfection of means and confusion of goals seem to ______ our age.
3) Men in the post office ______ mail according to places it is to go.
4) The throne symbolizing imperial power is positioned at the center of this central ______.
5) Protecting the earth our ______ is an duty for every citizen.
6) Their success does not necessarily ______ a leftward shift in politics.
7) She knows the ______ of the back streets like the back of her hand.
8) They heard voices coming from outside in the ______.
9) Columns are usually intended in architecture to add grandeur and ______.

Task 3 Challenge Yourself

3. Look at the picture below and talk about your understanding of the layout of Beijing four-in-one courtyard.

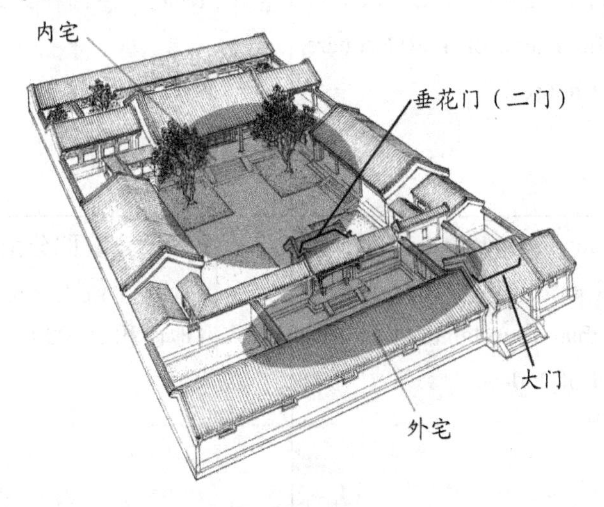

4. Translate the following paragraph into Chinese.

The characteristics of the courtyard houses in north China are front-hall-rear-room, the central axis being symmetric, the plane being regular and tidy, the exterior being prudent and enclosed, reflecting a profound hierarchical perception.

____

Task 4 Read and Simulate

5. The following English sentences on the left show some different ways of clarifying layouts. Read the English sentences carefully and translate the Chinese sentences on the right into English by simulating the structure of the English sentences.

Sample Sentences	Translation
1) The house originally built in 1875 is a typical courtyard residence in Tianjin, with a length of 96 meters from north to south, a width of 62 meters from east to west, covering an area of 6,080 square meters and a floor area of 2,945 square meters.	牟氏庄园坐落于栖霞县城北古镇都村。庄园坐北朝南，东西长158米，南北宽148米，分三组六院，共建厅堂楼房480余间。
2) The Princess Mansion was built in the 36th year of Emperor Kangxi of the Qing Dynasty (Year 1697), covering an area of 1.8 hectares and a floor area of 4,800 square meters, a total of 69 houses.	王家大院位于山西省晋中市灵石县的静升村，总计有大小院落123座、各种房屋1 118间，总面积约25万平方米。
3) The Mansion faces south and is divided into three layouts: east, west and center, consisting of more than four hundred houses and covering an area of 240-*mu*.	马氏庄园建筑群分南区、中区、北区，共6组，22个院落。建筑面积5 000平方米，占地面积20 000平方米。

Text B

Some Typical Chinese Traditional Courtyard Houses

The Four-in-one Courtyard of Beijing

Apart from the north courtyards' common characteristics of front-hall-rear-room, the central axis being symmetric, courtyards being interlinked with each other, the artistic image in the architectures of the Beijing four-in-one courtyard (北京四合院) is mainly reflected in the hall, the entrance gate, the secondary gate, the spatial texture inside the homestead, the greening of the court and the interior decoration and finishing.

The main entrance is the starting point of a whole four-in-one courtyard, and also the focal point of the artistic expression of the whole architecture's external form. Therefore, the shape and structure, and the classification of the main entrance became the hierarchical characterization of the whole set of buildings, and the symbol of the family's social status as well. Apart from enlarging the scale of the main entrance, the owner of a homestead adds such auxiliaries as the decoration on doors, the *Pushou* (the iron plates fixed on the doors to keep the knockers), the golden door-knockers, the door clasps, complemented with the stone drums, stone animals and even the engraved door stone/crossties outside the main entrance.

The entrance into the interior courtyard is called the floral-pendant gate, also the secondary gate, which is the focal point of decoration in the four-in-one courtyard and generally decorated magnificently.

The space composition of the four-in-one courtyard makes full use of the veranda, the wall to form a passing corridor by means of spacing or connecting. A small number of trees or miniature trees and rockery may be planted in the courtyard so as to create an atmosphere of quietness, easiness and comfort.

The Residence of the Shi Family, Tianjin

Located in Yangliuqing (杨柳青) Town of Tianjin City, the Residence of the Shi Family (石家大院) is the house of Shi Yuanshi (石元士), the fourth child of Shi Wancheng (石万程), one of the eight grand families in Tianjin in the Qing Dynasty. Renowned as "Zun Mei Tang (Respect & Beauty Hall, 尊美堂)", the house originally built in 1875 is a typical courtyard residence in Tianjin, with a length of 96 meters from

north to south, a width of 62 meters from east to west, covering an area of 6,080 square meters and a floor area of 2,945 square meters.

The general layout of the house is composed of the east yard, the west yard and the side yard.

The east yard is for living, consisting of three sets of courtyards. The principal room of each courtyard is in five-room breadth with a passage way in the middle. The wing-rooms in the east and the west are of three-room breadth. The west yard is a hall courtyard connected by two winding corridor yards, which is the most magnificent and the most beautifully decorated complex in the residence.

The south part of the house is formed by two front corridors of blue tile, hard top and paraboloid top. The front part serves as a hall for guests reception and banquets and the rear part is a theater.

Mansion of Princess Ke Jing, Hohhot, Inner Mongolia

The Mansion of Gu Lun Princess Ke Jing (固伦恪靖公主府) is located in Hohhot, Inner Mongolia, in the north of Guihua City in the old days. It was a time of turmoil in Khalkha-Mongolia when Gu Lun Princess Ke Jing got married, but Guihua City then enjoyed a time of stability and commercial prosperity. As an important north-south material distribution center in the desert and with the geographical advantage of being close to Beijing, Guihua City became the first choice to build the mansion.

The princess mansion was built in the **36**th year of Emperor Kangxi of the Qing Dynasty (Year **1697**), covering an area of **1.8** hectares and a floor area of **4,800** square meters, a total of **69** houses. It was larger than that of the Governor's Mansion in Guihua City at that time.

As an imperial residence, the mansion was constructed according to the standard prescribed by the Ministry of Construction of the Qing Dynasty, taking the great green mountain as the screen and lying in the arms of Zhadahai River and Aibugai River. Its style is very similar to the palaces in the capital city of Beijing which were built at the end of the Ming Dynasty and the beginning of the Qing Dynasty, adopting the traditional central axis of symmetry in architectural pattern in architectural system in the ancient China, constructing a large area of foundation and boasting hard-mountain features.

The mansion is a four-hall and five-yard residence, with a screen wall and a road for imperial carriage in the front and in the rear, garden, racecourse, house gate, ceremonial gate, providence hall, living room, wing rooms, backside house are arranged in hierarchical order. Till today, it is not only the most well-preserved princess mansion in the Qing Dynasty, but also the most typical courtyard building in that period.

Confucius Mansion in Qufu, Shandong Province

Queli Saint's Residence（阙里圣人家）— the Confucius Mansion（孔府）is located in Donghuamen Street of Qufu, Shandong Province. The mansion faces south and is divided into three layouts: east, west and center, consisting of more than four hundred houses and covering an area of 240-*mu*. The central part is the main body of the architecture, consisting of nine courtyards with offices in the front and residential quarters in the back.

The gateway of the Confucius Mansion has three doors and six door panels, and so is the second gateway. Within the second gateway, there grow high cypress trees and lush bushes. After the second gateway and through the Special Glory Gate, comes to the largest hall, which is followed by the second largest hall. These two halls are connected through a veranda, displaying a 工-shaped plane. At the left and right sides of halls are chambers. After three halls and six chambers are the inner living quarters. Its layout fully embodies the architectural features in the ancient China, government offices in the front and inner quarters in the back.

The inner quarters are living places for family members, including front main room, front chamber, back chamber and family chamber for worshipping Buddha, etc. Chambers are of two-floor architecture, with winding corridors and railings on the second floor. The design for the inner quarters is simple, elegant, grandeur and practical. At the end of the central part is the back garden.

Qiaojia Compound, Qixian County, Shanxi Province

As a castle-typed architecture, the Qiaojia Compound（乔家大院）was originally built during Emperor Qianlong's reign of the Qing Dynasty. With the enclosed high walls all around the courtyard, parapet on top, and with the watchtower and the balcony, the courtyard looks momentum. The entrance of the compound faces east and the opposite side of it is a shadow wall with brick-carvings of the scroll consisting of hundred forms of the character for longevity. Inside the main entrance is a stone paved path leading to the clan hall at the end of the courtyard.

The whole compound consists of six courtyards, three of them on each side of the road. The three courtyards on the north are typical of the local character of "the hallway courtyard", i. e. there are five principal rooms and five wing-rooms in the interior courtyard respectively but only three wing-rooms in the exterior courtyard, with a hallway connecting the two yards. The three courtyards on the south are all "double four-in-one courtyards". Each of the six courtyards consists of three-to-five small yards, with small yards in a larger one and one yard linked with another. The rooftops of all the houses in all the yards are connected with footpaths convenient for night watch.

Ingeniously designed and beautifully constructed with specifications and variety, the Qiaojia Compound has the aesthetic perception as a whole while each part has some unique features of its own. Whether the brick-carvings, the wood-carvings or the bucket arches, upturned eaves, doors and windows, etc. are all of different forms and the most varying.

Vernacular Architecture of Dangjia Village, Hancheng County, Shaanxi Province

Dangjia Village (党家村) was founded in the second year of the Zhishun Emperor's reign in the Yuan Dynasty (1331) with three times of large-scaled construction afterward.

The village is situated in the valley of the loess plateau, with the Mishui River passing around the south of it. Situated at the foot of the plateau and beside the river, the village is not only protected against cold wind and gets adequate sunshine, but also is able to drain water in accordance with the topography. That's why it is not very windy and dusty in the village though situated in the loess area.

Dangjia Village boasts itself for its scenic beauty and tile-roofed houses. The trees make a pleasant shade. The village, the stone tablets, the clan halls and the vernacular houses add radiance and beauty to each other. The bulwark of the village and the precipitous cliff are more than 30 meters higher than the ground of it, looking extremely magnificent. The upright-standing belvedere and the stone tablets in the village form the beautiful outline of the horizon.

The pavement of the village streets is made of stone blocks. With the T-shape, the lanes and the gates of the village do not face each other and the gates do not face the entrances of the alleys. There are 25 sentry posts guarding against theft at the entrance to the village, contributing greatly to the defense.

The gate-building of Dangjia Village is tall and big, with exquisite wood carvings and horizontal inscribed boards on the lintels and the steps on both sides of the gate and the ridges of the houses all have elegant brick-carving decorations.

Ma-clan Manor, Anyang City, Henan Province

The Ma-clan Manor (马氏庄园) is located in Xijiang Village, Jiangcun Town, Anyang City, Henan Province. It was a large-scale residence of Ma Piyao (马丕瑶), Deputy Minister of Defense (兵部侍郎) in the late Qing Dynasty, Deputy Surveillance Commissioner of the Ministry of Supervision (都察院右副都御史) and Governor of Guangxi and Guangdong (广西、广东巡抚). The architectural complex is divided into south region, central region and north region, with 6 groups and 22 yards. It covers an area of 20,000 square meters with a floor area of 5,000 square meters. As the largest official residence in the Qing Dynasty in Central China, it boasts 308 doors, wing-rooms, halls, corridors and buildings.

The original manor has been largely preserved. The north region, a large two-hall courtyard, is the ancestral residence and all the buildings are looking northwards. The central region, the largest compound in the manor, was built during the reign of Emperor Guangxu of the Qing Dynasty. It is composed of four courtyards sitting side by side from east to west and along the east-west orientation road in front of the door and all the four courtyards face south. The locals used to call the "yard route" which lies along the vertical axis in the traditional architecture. The west region is a courtyard with one main yard in the middle and two secondary yards on both sides. The layout of each courtyard is similar to each other and each is composed of a four-hall yard. The breadth of both the east and west yards is smaller to that of the main yard. The two-hall courtyard on the far east side is the ancestral hall and the family school of the Ma clan. The south region was built in the early period of the

People's Republic of China and its pattern and style are similar to that of those buildings in the central region.

Speaking and Writing

1. Work in pairs.

Suppose one is a tourist and the other is a tour guide. The guide is supposed to introduce the location, history and the layout of Qiaojia Compound, Confucius Mansion in Qufu, Mansion of Princess Ke Jing, Dangjia Village, and Ma-clan Manor respectively by using the pattern learned in Text A.

____

____

____

____

____

2. Summing up.

1) Write down what you have learned about Chinese traditional courtyard houses.

____

____

____

____

2) From this unit you have also learned

- useful words: ______________________________

____

- phrasal verbs: ______________________________

____

- useful expressions: __________________________

____

Unit 7

Crisscrossed-wood Structure

▶ Pre-reading Activities

1. Find out the common points of the buildings above in the four pictures and try to describe the structure of the first picture in your own words.
2. Given the climate and surrounding landform, why is the lower part of houses supported by wood poles?
3. Guess the usage of each storey.

Text A

Introduction to Crisscrossed-wood Structure

Many minority nationalities of our country live in the mountainous areas of the southwest, among which many places are hillside land teemed with wood and stone and so they have built their houses along slopes that are half-seated and half-hung in the air with wood columns sustaining the overhanging parts. Such houses are called overhanging houses of two or three stories with wood frame, wood floor, wood stairs and crisscross-wood-pole structure, in which livestock are raised on the ground floor while people dwell on the second and third floor. The rear of an overhanging house is situated on a slope with a village road in front of it leading to highways. Some slopes are rather steep, so the local villagers prop up the big wood houses under big eaves with long but thin posts, presenting a breathtaking and steady outward appearance with roughness and simplicity. The overhanging houses of the Dong people, the Miao people and the Zhuang people are typical examples among such traditional vernacular architectures of China.

The nationalities of the Dong, the Miao and the Zhuang mainly dwell on the boundary areas of Guizhou, Guangxi and Hunan Provinces, where the relief of the land is high and precipitous, the forests dense and thick, with a crisscross network of rivers and streams, and a damp and hot climate. People live in big or small stockades with their kinsmen, the layout of which has no rules and most houses are built at the foot of a hill, stretching freely according to the landform. The houses are detached wood ones with several stories. They are closely connected with each other with a narrow and long walkway in between. However, there must be a big opening serving as the place for public activities. In a Dong people's stockade, a huge multi-storied upright building called the drum tower is built on the opening and used as the center for the people in the stockade or of the same clan to gather together or discuss social, political and cultural affairs. As an important symbol of

the Dong people's residence, the drum tower is also the center of cultural entertainment, representing the Dong people's culture, and that is why it is called "drum-tower culture", too.

A rain-proof bridge is a transportation construction over a river or a stream, which is not only the passageway across the river, a space for inhabitants of a same clan to take a rest and communicate with each other at usual times but also the place for them to sing songs, drink alcohol and play *lusheng* (a reed-pipe wind instrument). It is located at the entrance of the stockade as the symbol of the entry to the stockade.

The houses of the Dong, the Miao and the Zhuang people are three-storied, crisscross-wood-pole buildings with a through-jointed frame, which are called overhanging houses by the local people. Two stories on the top are usually overhung from the sub-framing supported floor or overhung outward floor by floor. The house has a side corridor, a *xuanshan* roof (Chinese overhung gable-end roof) covered with tiles or fir skin, with a gentle inclination and far-reaching eaves. The overhanging houses look huge upward but small downward, the upper part being void while the lower solid. With simple and natural decorations, but without paintwork, an overhanging house presents lissome, vivid, flexible and unaffected style and features.

New Words and Expressions

teem /tiːm/	*v.*	crowded with people and/or animals 充满，充斥（人或动物）
precipitous /prɪˈsɪpɪtəs/	*adj.*	steep and often very dangerous 陡峭的
void /vɔɪd/	*adj.*	empty 空的
lissome /ˈlɪsəm/	*adj.*	gracefully slender; moving with ease 柔软的，轻盈的
crisscross-wood-pole	呈十字形的木制柱子，吊脚（楼）	
minority nationalities	少数民族	
drum tower	鼓楼	
rain-proof bridge	风雨桥	

Exercises

Task 1 Skimming and Scanning

1. Answer the following questions with the information contained in Text A.

1) In which part of China can you find crisscross-wood structure dwellings?
2) What geographic environment do these places have in common?
3) Which minority nationality usually dwell in drum tower?
4) What impressions would these structures have on people? Please describe the style of them.

Task 2 Language Focus

2. Fill in the blanks with the given words. You may not use any of the words in the bank more than once. Change the form of the given words if necessary.

teem	situate	prop up	vernacular	precipitous
detached	inhabitant	lissome	dwell	stretch

1) Our town has ten thousand ______ according to last year's statistics.
2) For most of the year, Yellowstone Park is ______ with tourists.
3) Traveling around can get you in touch with alien ______ buildings, some of which are breathtakingly stunning.
4) The ______ canoe seemed to be a fish, so easily did it cut through the rolling black waves and ranks of ice.
5) He had burns that ______ from his neck to his hips.
6) Camp sites are usually ______ along the coast, close to beaches.
7) They are concerned for the fate of the forest and the Indians who ______ in it.
8) The town is perched on the edge of a steep, ______ cliff.
9) The government does not intend to ______ declining industries.
10) In many small towns in America, you can see rows of ______ houses with beautiful gardens.

Task 3 Challenge Yourself

3. Think about the vernacular buildings in your hometown and talk about how can they display local residents' wisdom.

4. Translate the following paragraph into Chinese.

Such houses are called overhanging houses of two or three stories with wood frame, wood floor, wood stairs and crisscross-wood-pole structure, in which livestock are raised on the ground floor while people dwell on the second and third floor. The rear of an overhanging house is situated on a slope with a village road in front of it leading to highways. Some slopes are rather steep, so the local villagers prop up the big wood houses under big eaves with long but thin posts, presenting a breathtaking and steady outward appearance with roughness and simplicity.

Text B

The Vernacular Houses of the Tujia People in West Hunan Province

Most of the Tujia people live in the mountain area bordering of Hunan, Hubei, Sichuan and Guizhou Provinces. Under the condition of excessive rain, fog and humidity, they build a kind of overhanging houses with wood-frame, crisscross-wood-pole structure in line with the local conditions and materials, in which livestock are raised on the ground floor while people dwell on the second and third floor so as to avoid living in damp places and protect themselves from mountain torrents.

The layout of the houses is treated flexibly based on the topography and the needs of the functions. The houses may be built at the foot of a hill or by a stream, may face the sun or built in clusters, cohere with the natural environment to fuse the artificial beauty and the natural landscape into a whole. The method of building houses in layers along the slope is used according to the falling gradient. The ridges are deployed either along the parallel or vertical contour lines. Some roofs are built with the same height while some others are built downward layer upon layer along the terrain. To build houses on complicated landforms such as by a steep cliff of a stream, the method of half-hanging the houses in the air with wood columns sustaining the overhanging parts is used to gain more space for use.

The varied and gracefully-shaped houses form the unique style of the vernacular architecture of the Tujia people. In the treatment of eaves, the overhanging houses adopt the structure of back-curl corners and arc abat-vent, displaying a beautiful and elegant taste with its contour.

New Words and Expressions

humidity /hjuːˈmɪdəti/	*n.*	the condition when air feels heavy and damp 潮湿
torrent /tɔːrənt/	*n.*	a lot of water falling or flowing rapidly or violently 急流，洪流
topography /təˈpɑːgrəfi/	*n.*	physical shape of an area, including its hills, valleys and rivers 地形，地势
fuse /fjuːz/	*v.*	join sth. together 融合，融化
gradient /ˈgreɪdiənt/	*n.*	a slope, or the degree to which the ground slopes 斜坡，坡度
in line with	和……一致	
contour line	等高线	
back-curl corner	屋角反翘	
arc abat-vent	屋面举折	

Speaking and Writing

1. **Suppose you are Tujia residents and have lived in a vernacular house since born. Please introduce it on a TV programme in order to encourage more tourists coming to your hometown.**

2. **From this unit you have learned**
   - useful words: ____________________
   - phrasal verbs: ____________________
   - useful expressions: ____________________

3. **According to the text and other information about Tujia people, write an essay about 300 words describing the daily life of them in the third person. You can specifically focus on one fabricated person or describe the general life of common people.**

Unit 8

Vernacular Houses in the Region of Rivers and Lakes

▶ **Pre-reading Activities**

1. Look at the pictures above and tell us what is the color of walls and discuss why is that.
2. Have you ever been to Jiangsu or Zhejiang Provinces? What is people's general impression on the regions beside rivers and lakes and the girls' appearance and personality?
3. Can you recall a movie set in riverside or lakeside regions? Please briefly describe it.

Text A

Introduction to Vernacular Houses in the Region of Rivers and Lakes

The south of China has plenty of rivers, lakes and streams and numerous villages and towns are built along a river or beside a stream. The layout of the region of rivers and lakes consists of two main types: one type is the village that mainly functions as the residence while the other is the complex that functions both as a trading center and a residence. A river may link up the whole village, run through it or even surround it before running away. Frequently, two rivers may join each other in a village or a town so that the transportation and trade may become more convenient. Some shops or stores are just set up on the streets along the river, which have an easy access for shopping from both the land and water.

In villages and towns, the ordinary people's houses — the traditional vernacular architectures are arranged quietly, neatly and in good order. There are roads or streets in front of the houses and waterways behind them. Boarding on boats immediately after leaving one's home is a special feature indicating the convenient transportation in the region of rivers and lakes. Therefore boats have become indispensible means of transportation in such regions. Passing round the water lanes by boat is enough to experience and be attached to the boundless pleasure and appeal of the life in the region of rivers and lakes.

In a village, a bridge is a necessary transport component to link different parts over the stream-way there. It looks like a cobweb by linking in length and breadth as a fine landscape of the village. Bridges such as straight bridges, level bridges, arch bridges, gallery bridges, circular-orifice bridges, stair-step bridges and treasury-belt bridges, and so on, add limitless fine landscape to the villages and towns in the region of rivers and lakes. In addition, some towers or pavilions are also built at the head of the bridges where people can take a rest while some shops, stores or restaurants are built beside the rivers, presenting a flourishing sight of the region's construction.

Seen from a distance, the adjoining black-paint gates, the homesteads with grey tiles and white wall situated along the slab-stone paths, along with the poplar trees, the willow trees, the fruit trees and the bending weeping willows, the stretch of wharfs behind them, several small boats drifting on the water of the stream, look together really like an easy, comfortable, peaceful and secluded picture in which people in the region of rivers and lakes can live a life of leisure.

New Words and Expressions

cobweb /ˈkɑːbweb/ *n.* a net which a spider makes for catching insects 蜘蛛网

wharf /wɔːrf/ *n.* a platform by a river or the sea where ships can be tied up 码头

secluded /sɪˈkluːdɪd/ *adj.* being quiet and private 僻静的，隐蔽的

slab-stone 石板

Exercises

Task 1 Skimming and Scanning

1. Answer the following questions with the information contained in Text A.

1) How many types are consisted of the layout of the region of rivers and lakes? And what are they?

2) How are villages linked with one another?

3) Can you describe the scenery around rivers and lakes according to the text?

4) What is the impression these houses have on people?

Task 2 Language Focus

2. Fill in the blanks with the given words. You may not use any of the words in the

bank more than once. Change the form of the given words if necessary.

indispensable	secluded	adjoin	complex
layout	consist of	boundless	landscape

1) The geographical condition of our factory is superior, ______ the harbor of Tianjin.
2) May your fortune be as ______ as the East Sea and may you live a long and happy life!
3) With a conventional repayment mortgage, the repayments ______ both capital and interest.
4) He wants to add a huge sports ______ to Binfield Manor.
5) An intelligent computer will be a/an ______ diagnostic tool for doctors.
6) We meandered through a ______ of mountains, rivers, and vineyards.
7) The summer resort has a simple and elegant ______.
8) Altogether, it was a delightful town garden, peaceful and ______.

Task 3 Challenge Yourself

3. Please match the pictures below with their corresponding bridge type.

A. straight bridge B. arch bridge C. gallery bridge
D. circular-orifice bridge E. stair-step bridge F. treasury-belt bridges

1) ______

2) ______

3) _____ 4) _____

5) _____ 6) _____

4. Translate the following paragraph into Chinese.

Seen from a distance, the adjoining black-paint gates, the homesteads with grey tiles and white wall situated along the slab-stone paths, along with the poplar trees, the willow trees, the fruit trees and the bending weeping willows, the stretch of wharfs behind them, several small boats drifting on the water of the stream, look together really like an easy, comfortable, peaceful and secluded picture in which people in the region of rivers and lakes can live a life of leisure.

Text B

Tongli Town, the Region of Rivers and Lakes, Jiangsu Province

Tongli Town is situated in northwest Wujiang County, Jiangsu Province. Surrounded by water all around, the town has beautiful scenery. The whole town is divided into seven small islands by fifteen streams and forty-nine historic bridges fuse the seven islands into an organic whole. The streets and alleys meander within the town and rivers crisscross. Every house is close to water and every home is navigable. With deep lanes and quiet environment, the tilt-covered houses with white walls and colorful windows roll uninterrupted, and have the unique style and features of the region of rivers and lakes.

The beautiful environment, convenient transportation and rich produces in Tongli Town become the reasons for the gentry to go and live in seclusion and take care of themselves there. Therefore, there are plenty of compounds of connecting courtyards, each surrounded by dwelling quarters as well as excellent vernacular houses in the town. There are more than ten historic architectures built during the Ming Dynasty and several scores of houses built in the Qing Dynasty (such as the Tuisiyuan Garden) existing in the town, which are called the architectural museum of the Ming and Qing Dynasties.

The Tuisiyuan Garden is a very exquisite private garden built with originality in the Qing Dynasty. It consists of residences, front yards and gardens from the west to the east. The Tuisiyuan Garden is a special case among the gardens pressing close to water in the region south of the Yangtze River, in which the hills, pavilions, galleries, corridors and verandas all press to the water so that the Garden looks as if floating on the water. Water is the source of all lives and people are pleased to enjoy living close to water, striding over water, and being on intimate terms with water.

New Words and Expressions

meander /miˈændər/ *v.* (river or road) have a lot of bends, rather than going in a straight line from one place to another (河流、道路)蜿蜒而行，迂回曲折

navigable /ˈnævɪɡəbl/ *adj.* being wide and deep enough for a boat to travel along safely 可航行的，适合航行的

gentry /ˈdʒentri/ *n.* people of high social status or high birth 上流社会人士，绅士贵族阶层

originality /əˌrɪdʒəˈnæləti/ *n.* with great imagination and new ideas 独创性，新颖

on intimate terms with 与……关系密切

Speaking and Writing

1. Suppose you are the owner of Tuisiyuan Garden, and discussing the design details with craftmen. Please create a dialogue between you and them.

Words for reference: originality, exquisite, rockery, live in seclusion.

2. From this unit you have learned

- useful words: ______________________________

____

- phrasal verbs: ______________________________

____

- useful expressions: ______________________________

____

3. Collect information about riverside houses in Zhejiang Province, and write an essay about 300 words summarizing the change of them since ancient times.

Unit 9

Cave Dwellings

▶ Pre-reading Activities

1. Find out the common points of the buildings in the four pictures and try to describe the structure of the first picture in your own words.
2. Given the climate and surrounding landform, why is the cave surrounded by the hills or dug down into the ground?
3. Please guess the usage of each cave.

Text A

Introduction to Cave Dwellings

The Yellow River Valley in the Central Plains of China cuts across Gansu, Shaanxi, Shanxi, and Henan Provinces with particularly favorable natural conditions of the loess resources, the geography of which is well-distributed and spread in succession with a thickness of 50 to 200 meters. As the major ingredient of the loess is silt made up of quartz whose granules are relatively fine with a rather high viscosity, quite a strong stickiness and shear stiffness, it is easy to dig for construction. The cave dwelling, both cold-proof and warm, is one of the most suitable means of inhabitancy for the local ordinary people to draw on local resources for practical purposes.

The cave dwellings can be classified into three categories according to landforms. The first are hill-backed cave dwellings, which are dug horizontally directly into the hill or cliff. This category has two subdivisions: one is the hill-back type, in front of which there is a wide open space. The cave dwelling dug along the slope and the ground floor is also the platform of the upper cave dwelling. The other are called gully-type cave dwellings, which are distributed in the solum of the precipices along both sides of a gully. Because the gully is relatively narrow, the open space in front of the cave dwellings is small. However, due to the narrowness of the gully, the caves can easily avert wind and sand blow.

The second category are subsiding-type cave dwellings or small-courtyard cave dwellings. In the flat region where there is no condition for people to dig cave dwellings into the precipices, they have to dig down into the ground so as to build a closed subsiding courtyard with precipices on four sides and then make cave dwellings by digging into the precipices, which are called small-courtyard cave dwellings. Such cave dwellings are called "small courtyards" in Henan Province, "grotto courtyards" in Gansu Province, "basement courtyards" or "pit courtyards" in Shanxi Province. The passageway for the dwellers to come in and go out is the ladder.

The third category are the covering caves, also called detached cave dwellings, which in essence are a kind of arch buildings made of sun-dried mud bricks or masonry blocks, the upper part of which is tamped with earthing.

The mono-plane of a cave dwelling is a rectangular room with a vault on top and the two adjoining caves are opened up. The door is set in the frontage, namely, the face of the cave. The lower part of the frontage is the sill wall window while the upper part is the arch window beside which the gate is usually set. Some cave dwellings have windows but don't have doors, which are used as the pit beds. The outward appearance of a cave dwelling is concise and simple without making a display of it. Some decorations may be made on the door frame or lintel and window panes may have paper-cuts stuck on them. Fused together with the loess earth, the cave dwellings maintain the natural ecotope styles and features, leaving people a sense of natural beauty in the countryside with simplicity and roughness.

New Words and Expressions

silt /sɪlt/	*n.*	mud or clay or small rocks deposited by a river or lake 淤泥，泥沙
granule /ˈgrænjuːl/	*n.*	a tiny grain 颗粒
viscosity /vɪˈskɑːsəti/	*n.*	resistance of a liquid to sheer forces (and hence to flow) [物] 黏性，[物] 黏度
inhabitancy /ɪnˈhæbətənsi/	*n.*	the act of dwelling in or living permanently in a place (said of both animals and men) 住所；有人居住的状态
solum /ˈsoʊləm/	*n.*	the upper part of the soil profile, which is influenced by plant roots 土壤表层；风化层
avert /əˈvɜːrt/	*vt.*	prevent the occurrence of; prevent from happening 避免，防止
precipice /ˈpresɪpɪs/	*n.*	a very steep cliff 悬崖，绝壁
grotto /ˈgrɑːtoʊ/	*n.*	a small cave (usually with attractive features) 岩穴，洞穴；人工洞室
ecotope /ˈekətoʊp/	*n.*	ecological environment 生态环境
in essence	本质上	
stick on	保持在……上，贴上	

Exercises

Task 1 Skimming and Scanning

1. Answer the following questions with the information contained in Text A.

1) In which part of China can you find small-courtyard cave dwellings?

2) How many categories can the cave dwellings be classified into according to landforms? What are they?

3) What impression would the outward of the cave dwellings have on people?

Task 2 Language Focus

2. Fill in the blanks with the given words. You may not use any of the words in the bank more than once. Change the form of the given words if necessary.

silt	shear	ecotope	stuck on	avert
adjoining	ingredient	precipice	viscosity	essence

1) Competitors have six minutes to ______ four sheep.

2) Magnetic fields reduce blood ______.

3) The car backed through the wall after the driver's foot ______ the accelerator.

4) But remember, none of these recipes for success will work without the key ______: willpower.

5) This, in ______, is what companies most value from vendors.

6) The river deposited ______ at its mouth.

7) A disaster was narrowly ______.

8) Watching him climb up the ______, everybody was breathless with anxiety.

9) But to tell you the truth, he snores so loudly that people in ______ rooms have complained in the past.

10) The renewal of construction of ______, is the chief work to develop the west of China.

Task 3 Challenge Yourself

3. Think about the features of hill-backed cave dwellings, subsiding-type cave dwellings and detached cave dwellings, and talk about how they display local residents' wisdom.

4. Translate the following paragraph into Chinese.

The Yellow River Valley in the Central Plains of China cuts across Gansu, Shaanxi, Shanxi, and Henan Provinces with particularly favorable natural conditions of the loess resources, the geography of which is well-distributed and spread in succession with a thickness of 50 to 200 meters. As the major ingredient of the loess is silt made up of quartz whose granules are relatively fine with a rather high viscosity, quite a strong stickiness and shear stiffness, it is easy to dig for construction. The cave dwelling, both cold-proof and warm, is one of the most suitable means of inhabitancy for the local ordinary people to draw on local resources for practical purposes.

____

____

____

____

Text B

The Kang Baiwan's Manor of Gongyi, Henan Province

The Kang Baiwan's Manor（康百万庄园）faces the Yiluo River on the south with its back against the Mangshan Mountain, having a very peaceful and secluded environment. Its construction began during Emperor Daoguang's reign and was completed during Emperor Xuantong's reign of the Qing Dynasty, lasting for decades. The manor, a typical castle-cave residence in the area of the Loess Plateau, North China, consists of the ancestral temple, the residence of the owner of Jingu Stockade, average residences, workshops, warehouses, etc.

The residence of the owner of Jingu Stockade is composed of four juxtaposed four-in-one courtyards and a wing cave-courtyard made out of the cliff. Only the first courtyard has a main hall while the others all take the cliff caves laid with bricks as their main rooms respectively. The rooms in the first yard are well-proportioned and have symmetrical central axis. The second, third and the fourth yards all have a second gate with festoon curl canopies. In front of the second gate there is a four-meter wide east-to-west sidewalk connecting all the yards, and in the backyard a two-meter wide passageway runs from east to west of the yards. In this way, a huge resident with all the yards independent as well as mutually interlinked with one another is formed.

The Kang Baiwan's Manor is beautiful, imposing and sumptuous. Its exquisite and unique post and panel structure, straight and impressive-looking overlying-ridge gable walls, dainty and exquisite patterns on the doors and windows, refined and gorgeously carved beams and painted rafters, and the elegant and unsophisticated furniture, all add strong local characteristics to the manor.

New Words and Expressions

manor /ˈmænər/	*n.*	the mansion of a lord or wealthy person 庄园
juxtapose /ˌdʒʌkstəˈpoʊz/	*v.*	place side by side 并列；并置
well-proportioned	*adj.*	of pleasing proportions 均衡的；比例恰当的
symmetrical /sɪˈmetrɪkl/	*adj.*	having similarity in size, shape, and relative position of corresponding parts 匀称的，对称的
sumptuous /ˈsʌmptʃuəs/	*adj.*	rich and superior in quality 华丽的，豪华的；奢侈的
ancestral temple		祠堂；宗庙；宗祠

Speaking and Writing

1. Suppose you are the resident from Kang Baiwan's Manor of Gongyi and have lived in vernacular houses since born. Please introduce it on a TV programme in order to encourage more tourists coming to your hometown.

2. From this unit you have learned

- useful words: ______________________________
____
- phrasal verbs: ______________________________
____
- useful expressions: ______________________________
____

3. According to the text and other information about Manor people, please write an essay about 300 words describing the daily life of them in the third person. Compare and analyze the differences between Kang Baiwan's Manor of Gongyi and other cave dwellings.

Unit 10

Garden Houses

▶ Pre-reading Activities

1. Look at the pictures and tell us what the roofs look like and discuss why they are like that.
2. If you live in a garden house, what should it look like? Imagine and describe your ideal garden house.
3. Do you think the garden house must be built in the south? Is it influenced by climate?

Text A

Introduction to Garden Houses

A special kind of houses accommodating the Han people used to be inhabited by old and well-known families, literati and officialdom (in feudal China). Later, some ordinary citizens also lived in such houses. They are vernacular architectures called garden houses or garden court houses that combine residences, studies and garden courts together. Such houses are classified into three types as follows.

First, a study is built beside a residence and some detached studies have styles of their own as they are built independently. The plane feature of such a study is that in a three-bay house, the hall is extended forward into a long one. On both sides of the jut of the long hall, each small yard is arranged respectively. In the small yards, some slim bamboos, trees and grass are planted and a small amount of stones with odd shapes is banked up so that the courtyard is afforested on both sides of the study. Big windows are installed in both sides of the long hall. Such a study, though small, has a very quiet and peaceful surrounding.

Second, an independent garden with a style of its own, is built beside a residence, which is commonly seen in the region south of the Yangtze River and other regions of south China. The feature of its layout is mainly ornamental, without residence in it. The residence adjoins the garden opening onto each other with doors, but both the residence and the garden are independent.

Third, the residence, the study and the garden are grouped into one homestead, in which each one is independent with its own distinct function. It has such a plane feature as to take residence as the principal part and the study and the garden as subsidiaries. The environment within the homestead is quiet and one can read poems, verses aloud and savor painting and calligraphy in the study and stroll idly in the garden at his leisure.

In design of the garden, the hall is taken as the center, complemented with rockery, pond water, flowering wood, grass all around, so that each scenic spot will be formed by spacing the garden with a corridor, a wall, a bridge or a pavilion. It is required that on a limited piece of land, full use be made of the surrounding environment, the architectures and the natural landscape, and the means of the assembled view, the scenic focal point and the borrowed view be adopted so as to create more and richer sceneries to satisfy the garden owners' needs to enjoy the sight of the garden, to live in the garden and to hold parties or games in the garden.

The garden residence emphasizes the association of activity and inertia, focusing quietness and tastefulness. The book *Yuanye* by Ji Cheng in the Ming Dynasty pointed out that in the creative design of a garden, 30% belongs to the craftsmen while 70% belongs to the owner, which indicates that the owner plays the leading role in the design. The book also pointed out that the art lies in cleaver borrowing and the essence rests with suitability, which is the criterion for creating the design of the garden residence. Through the construction practice of the garden residence, one can not only obtain the artistic enjoyment of residence and play, but also savor the aesthetic value of it.

New Words and Expressions

subsidiary /səbˈsɪdieri/ *n.* an assistant subject to the authority or control of another 附属事物，附属机构

savor /ˈseɪvər/ *v.* derive or receive pleasure from; get enjoyment from; take pleasure in 尽情享受；品尝，欣赏

idly /ˈaɪdli/ *adv.* in an idle manner 无所事事地；空闲地

literati and officialdom 士大夫

Exercises

Task 1 Skimming and Scanning

1. Answer the following questions with the information contained in Text A.

1) How many types are the garden houses classified into? And what are they?

2) What are the means adopted in design of the garden? And what's the purpose?

3) Who inhabited in the garden houses?

Task 2 Language Focus

2. Fill in the blanks with the given words. You may not use any of the words in the bank more than once. Change the form of the given words if necessary.

bank up	ornamental	afforest	savor
pavilion	idly	subsidiary	odd

1) In compact space, it still deserves to have reasonable ______.
2) The marketing department has always played a ______ role in the sales department.
3) I'm wearing ______ socks today by the way.
4) They may sit ______ for hours, and stop eating, bathing, going out — even getting out of bed.
5) ______ the joy of simple pleasures.
6) "Don't think you can just ______ your sleep on the weekend, because it doesn't work that way," Harris warned.
7) The ______ on the hill looks down on the river.
8) They are not only ______, but also useful.

Task 3 Challenge Yourself

3. The following is a list of terms. After reading it, you are required to find the items equivalent to those given in Chinese in the list below. Then you should put the corresponding letters in brackets.

A. stroll idly
B. homestead
C. cave dwellings
D. artistic enjoyment
E. pavilion
F. garden court houses
G. savor painting and calligraphy
H. savor the aesthetic value
I. garden residence
J. study
K. courtyard houses
L. essence rests
M. corridor

1) ( ) 园林民居
2) ( ) 住宅
3) ( ) 朗读诗词
4) ( ) 艺术享受
5) ( ) 走廊
6) ( ) 窑洞民居
7) ( ) 书斋
8) ( ) 悠闲漫步
9) ( ) 品味美学价值
10) ( ) 亭子

N. rockery

O. residence

P. read poems, verses aloud

4. Translate the following paragraph into Chinese.

A special kind of houses accommodating the Han people used to be inhabited by old and well-known families, literati and officialdom (in feudal China). Later, some ordinary citizens also lived in such houses. They are vernacular architectures called garden houses or garden court houses that combine residences, studies and garden courts together.

____

____

____

____

Text B

The Vernacular House Gardens in Suzhou, Jiangsu Province

The private gardens of Suzhou were in fashion during the Ming and Qing Dynasties, most of which were not large since they were just a part of the homesteads. The flexible and varied techniques of treating the gardens' spaces, the peculiar and exquisite rock-piling and water feature design, the just-perfect plants concoction, the ingenious and elegant architecture complex, the pure, honest and rich cultural connotation, all lead people to endless aftertastes.

Without water, there would be no gardens because water is the blood vessel of gardens and gives gardens limitless vitality. Gardeners took effective measures according to the local conditions, made ponds from the level ground, and piled up a mount with earth. The mounts and water were properly distributed with distinct gradations.

Water makes people think of far away and rocks help people recall the ancient. Mounts are the skeletons of gardens. The piled rocks and earth with ridges, peaks, gullies, abodes of fairies and immortals, and precipices in the gardens seem to have the veins and imposing manners of the real mountains and cause gardens to have endless variations. The materials selected for constructing rockeries in gardens are mainly rocks from the Lake Tai because fine rocks from the lake have the four features: creasy, thin, leaky and transparent. By being creasy, the rocks must have twists and turns, gradations and variations. By being thin, the rocks must be tall, straight, delicate and pretty. By being leaky, the rocks must be interesting and appealing. By being transparent, the rocks must be dainty and exquisite.

The house gardens in Suzhou draw the essence of the folk architecture in the region south of the Yangtze River, light, handy and slender, ingeniously and delicately wrought, displaying a quiet, elegant and tasteful scenery as both the interior and exterior of the houses with white-washed walls and grey tiles are well-ventilated, and the scenery in the garden is brilliant with hills and waters, flowers and trees.

New Words and Expressions

concoction /kən'kɑːkʃn/	*n.*	something that has been made out of several things mixed together 混合物；调合物
aftertaste /'æftərteɪst/	*n.*	a taste that remains in your mouth after you have finished eating or drinking something 回味；余韵
skeleton /'skelɪtn/	*n.*	frame of bones in one's body 骷髅；[解剖]骨骼

Speaking and Writing

1. **Suppose you are a tourist guide, please introduce the features about the vernacular house gardens in Suzhou to the tourists.**

   Words for reference: homesteads, peculiar and exquisite, mounts and water, essence, aftertaste.

2. **From this unit you have learned**
   - useful words: ______________________________
   ____
   - phrasal verbs: ______________________________
   ____
   - useful expressions: ______________________________
   ____

3. **What are the differences between the vernacular house gardens and houses in the region of rivers and lakes in Jiangsu Province? Please collect some information about them and write an essay about 300 words summarizing the differences between them since ancient times.**

参考答案及译文

Unit 1 The History and Culture of the Chinese Traditional Vernacular Architecture (中国民居建筑的历史与文化)

【参考答案】

Text A

Task 1

1. Answer the following questions with the information contained in Text A.

1) Clothes, food, accommodation and transportation are the four basic necessities of man's life.

2) Because in the South, the climate is hot and humid and there are insects and wild animals in the open country.

3) The small houses in the Han Dynasty were usually of square or rectangle shape.

4) In the Song Dynasty, the main part of the residence is a 工-shape house consisting of the front hall, corridor and rooms at rear; the 工-shape or 王-shape plane is connected by a corridor in the middle.

2. Text A can be divided into three parts with the paragraph number(s) of each part provided as follows. Write down the main idea of each part.

Part One: the origin of the Chinese vernacular architecture

Part Two: the Chinese vernacular architecture in the primitive society

Part Three: the Chinese vernacular architecture in the feudal society

Task 2

3. Match each housing architecture with the period it first appeared.

courtyard—Sui and Tang Dynasties

cave living—Primitive people

garden house—Northern and Eastern Wei periods

bole-fence house—Hemudu people

工-shape house—Song Dynasty

4. Fill in the blanks with the given words. You may not use any of the words in the bank

more than once. Change the form of the given words if necessary.

1) dwell 2) excavate 3) access 4) habitat 5) priority
6) distinction 7) characteristics 8) surge 9) prototype 10) stretch

Task 3

5. Translate the following paragraph into Chinese.

明清时期，人口骤增，经济兴旺、文化发展，城乡面貌更显繁荣。该时期大量的民居建筑实物至今尚存，遍布各地，基本上仍保留着各地各民族民居的传统特色。

【参考译文】

Text A 中国民居建筑的起源与发展

衣食住行是人类生存的四大基本需求。为了满足住的需要，于是产生了居住建筑，它是历史上最早出现的建筑类型。

远古时期，原始人居住在天然洞穴之中，称为“穴居”。在南方，因气候湿热，野外多虫兽，于是在树上构筑简陋的窝棚栖息，称为“橧巢”，也称“巢居”。其后居住建筑发展到半地下，再到地面，在我国陕西半坡有大量的遗址为证。

在长江下游，浙江余姚河姆渡发掘出的聚落遗址，是一座长度在30米以上的干栏式长屋，进深约7米，前檐宽约1.3米，建于木桩上。长屋地板比地面高约1米，用木梯上下。其梁柱等构件交接处的榫卯构造已相当复杂，可见当时木构建筑技术已有很大进步。

封建社会早期以农业生产为主。从墓中出土的画像石、画像砖和用具、陶屋中，可以推知汉代小型住宅的平面为方形、长方形，稍大的住房有曲尺形，内部有院落，正中有楼高起，次屋则低矮，外观主次分明。贵族的大型宅第有几进院落，正屋旁还有杂屋、宾客住房。院落内前堂为主要建筑，后堂有屋，是古代前堂后室制的发展。

从北魏东魏时期留下的石刻中可以看出，贵族住宅有大型厅堂和庭院回廊，甚至在住宅后部建有园林，是园林式住宅的最初原型。

隋唐五代的民居建筑，仅能从敦煌壁画和其他绘画中得到一些旁证，如贵族宅邸大门采用乌头门形式，宅内两座主要房屋之间用带有直棂窗的回廊，连接成院。

宋代农村住宅可见于宋画《清明上河图》中，茅屋比较简陋，墙身较矮，有些是茅屋和瓦屋相结合构成的一组房屋。城镇小型住宅多用长方形平面，屋顶为悬山式或歇山式，除草葺和瓦葺外，山面的两厦和正面的披檐则多用竹篷或屋顶上加建天窗。稍大的住宅，外建门屋，内部采用四合院形式，有些在院内栽花植树，美化环境。

此外，宋朝王希孟《千里江山图》卷中所绘住宅多所，都有大门、东西厢房，而主要部分是前厅、穿廊、后寝所组成的工字屋。这种在古代前堂后寝制的传统布局原则下，中间用穿廊连成工字形、王字形平面，成为本时期民宅布局的一大特点。

明清时期，人口骤增，经济兴旺、文化发展，城乡面貌更显繁荣。该时期有大量的民居建筑实物至今尚存，遍布各地，基本上仍保留着各地各民族民居的传统特色。

Text B 中国民居建筑与文化

中国历史悠久，土地辽阔，文化遗产极为丰富。作为建筑文化遗产中的数量最大的，与广大人民生产、生活密切相关的传统民居，也同样非常丰富。今天，它们仍然散布在全国各民族和各地区之中，虽经风雨沧桑，仍然为广大人民所使用。其中不乏传统精华，它们艺术精湛，是我国珍贵的文化宝藏。

我国南北气候悬殊，山丘河海地理条件各不相同，材料资源又有很大差别，加上各民族不同的风俗习惯、生活方式和审美要求，导致我国民居建筑具有鲜明的民族特色和丰富的地方特性。

中国传统民居建筑与社会、历史、文化、民族、民俗有关，又与儒礼、道学、阴阳五行等思想学说有密切联系。优秀的传统民居建筑，具有历史文化价值；同时，又有实用和艺术价值，这是我国宝贵的文化遗产，也是世界文化的珍贵财富，亟须保护和弘扬。

哲学是文化思想的集中体现，是民族精神的精华，是人类智慧的最高创造。在中国传统文化系统中，哲学处于核心地位，起着主导的作用。

从哲学看，中国传统文化主要受下列思想影响，即儒学、道学、阴阳五行学，以及其后的中国佛学。它们相互影响吸收融合，构成了中国传统文化的总体。

中国传统文化的基本精神涵盖四个主要方面，即以人为本的人文主义价值系统，自强不息、豁达乐观的民族心理，观物取象、整体直觉的思维方式，以及天人合一的审美理想。就价值系统而言，中国文化表现了突出的以人为本的实用理性精神。

中国传统文化的思维方式有两个特点：直觉体悟的直观性和观物取象的象征性。天人合一是中国传统文化的审美理想和最高境界。

中国传统民居是中国传统文化在建筑中的体现。通过中国传统民居的布局、形制，我们可以形象而深刻地感受到中国传统文化思想的深远影响。中国传统民居充分体现了中国传统文化中的哲理观、宗法观、环境观、思维观，从不同层面反映了中国传统文化的渊博精深和高明智慧。

中国传统文化在中国民居建筑中的影响，首先是宗法观，即宗法礼制。宗法观的核心是等级观念。它对民宅中的布局、形制，甚至开间、规模、装饰、装修都作了严格的规定，目的是维护以血缘为纽带的封建等级制度。

其次是天人合一观念，它是民居择向、定位、选址的依据，又是汉族、儒家道家思想在大环境条件下处理建筑的准则，它包括建筑平面布局和空间组织结构的整体性、秩序性和教化性。在古代，中国传统民居天人合一的环境理想大多通过风水论说来实现，这就是五位四灵的环境模式，五位即东、南、西、北、中五个方位，四灵即四方神灵青龙、白虎、朱雀、玄武。

再次是中国传统民居中的思维观，它充分反映了中国传统文化中的人本主义精神，一切以人为本，以民居建筑中的主人及其家庭为本，天、地、人三者结合，因此把人的生产、生活作为民居营建的出发点。

这就是中国传统文化在中国传统民居中的影响和特征。

Unit 2 The Composition and Characteristics of the Chinese Traditional Vernacular Architecture
（中国民居建筑的构成和特征）

【参考答案】

Text A

Task 1

1. Answer the following questions with the information contained in Text A.

1） They have far-reaching impact on the vernacular houses in terms of the layout, room structure and the scale.

2） The hierarchy of the feudal system was so stern that it had hard and fast rules on the architecture in terms of the scale, size, width, spatial room, as well as the roof form, and even the furnishing, decoration and color.

3） The vernacular house is constructed and completed in a given location and environment and under given geographical and climatic conditions. The dry and cold weather in the north and the hot and humid weather in the south lead to different approaches and methods in constructing vernacular architectures in different places.

4） Such a layout is reasonable by the analysis of the modern concepts. For example, the running water lies in front of the village, meeting people's needs for fresh water, transport and washing; the high mountain behind it can be a perfect screen to resist cold wind; building a house on a tilting terrain keeps the house dry and makes it easy for drainage, which is ideal for living and health.

5） In accordance with the Five-Elements theory, water suppresses fire while gold gives birth to water. Thus, the purpose of water and gold gable is to check and prevent fire.

Task 2

2. Fill in the blanks with the given words. You may not use any of the words in the bank more than once. Change the form of the given words if necessary.

1） involve　2） resist　3） dominate　4） doctrine　5） impose
6） advocate　7） accordance　8） ethnic　9） indispensable
10） restriction

Task 3

3. Translate the following paragraph into Chinese.

风水观，古称堪舆学，它来源于阴阳五行学说，原是古代阳宅和阴宅分别择位定向时考虑气候、地理环境的一门学说。阳宅即民居建筑，例如，在农村，对民宅的选址一般已形成一种比较固定的负阴抱阳模式，即村前要有流水，村后要有高山，房屋坐北朝南，地形前低后高。从现代观念来分析，这种布局原则还是有科学性的一面。譬如村落面靠流水，这是食水、交通、洗濯的需要。村后高山作屏，可抵御寒风侵袭。地形前低后高，说明坡地上盖房子既要求干燥又要易排水，对居住及人体健康有益。

【参考译文】

Text A　中国民居建筑的构成

中国民居建筑的构成有四个因素，即社会因素、经济因素、自然因素和人文因素。

社会因素，包括生产力、社会意识、民族差异、宗教信仰和风俗习惯等

我国是一个多民族国家。以汉族为例，在历史上长期是一个以宗法制度为主的封建社会，家庭经济则以自给自足的小农生产为基础，并以血缘纽带作为宗族的维系，而维持社会稳定的精神支柱则是儒家伦理道德学说。这种学说提倡长幼有序、兄弟和睦、男尊女卑、内外有别等道德观念，并崇尚几代同堂大家庭共同生活，以此作为家庭兴旺的标志。对民居建筑来说，它对内要满足生活和家庭生产的需要，对外则要防止干扰，实行自我封闭，尤其对妇女的活动严格限制在深宅内院之中。宗法制度的另一重要内容则是崇祖祀神，提倡对宗族祖先的崇拜和对各种地方神祇的祭祀。这种宗法制度和道德观念对民居的平面布局、房间构成和规模大小有着深远的影响。

中国封建制度的核心是等级制度和儒礼宗族制。汉族院落式的民居平面布局就是这种形制的产物。封建制度等级森严，无论建筑的规模、大小、开间、进深以及屋顶形式，甚至装饰、装修、色彩都有严格规定。例如民宅不得超过三间，色彩规定黑白素色。而大型宅第就可以多区域、多院落，甚至多路建筑布局，并且可以带书斋、园林。从民居的平面布局就可以看到社会制度对建筑的影响。

经济因素是民居形成的物质基础

民居的营建需要材料，并以一定的构造方式建造起来。因此，宅居所用材料的数量、品质和结构方式决定着民居建筑的规模、质量和等级。富有者可在建筑的大门、屋顶和室内进行华丽的装修，而贫穷者只能以泥墙挡风雨、薄瓦以蔽身。

然而，劳动人民的智慧是无限的。他们利用当地的材料如木、竹、灰、石、黄土，根据当地的自然条件和自己的经济水平，因地制宜，因材致用，按照自己居住的需要和营造的规律来进行建造。因此，他们的民居具有实际、合理的功能，设计灵活，材料构造经济，外形简约，体现建筑的本质功能。特别是，广大民居建造者和使用者是同一的，是自己设计、自己营建、自己使用三位一体，因而，民居更富有人性、经济性和现实性，也最能反映本民族的特征和本地区的地方特色。

自然因素，包括气候、地形、地貌、材料等自然物质和环境因素

我国东南海域辽阔，西北高山峻岭。南北总长约5 500公里，东西宽度约5 200公里。境内地貌西北高、东南低，西部有广阔的高原，东部有大片平原，其间分布着高山、丘陵，东南则多水网密布的河湖和溪流。

我国气候南北差异很大：北部寒冷，并有霜雪；南方地区夏季炎热、潮湿、多雨，有的地区终年不下雪。每年夏秋季，东南沿海地区常有台风侵袭，时常带有暴雨，对人畜、房屋伤害极大。

此外，由于地理、气候的不同，各地制作建筑材料的自然资源也有很大差别。中原及西北地区多黄土，丘陵山区多产木材和石材，南方还盛产竹材和砖块，许多地区都能自产砖瓦和采集天然的沙石，沿海还能生产由贝壳烧成的蜃灰。

民居建造是在一定的地点、环境，一定的地理、气候条件下完成的。北方的天气干燥、寒冷，南方的气候闷热、潮湿，导致南北民居建筑的处理方式、手法都不一样。以地理环境来说，有坡地、平地、河流、小溪、山。民居建在坡地或平地上，或建于水畔，其景观效果都不一样。特别是气候因素对民居建筑的平面布局、建筑造型以及内部空间影响更大，这也是不同地区民居呈现不同形态、不同特色的重要原因。

人文因素，包括民情、民俗、生产、生活方式以及文化、审美观念

汉族民居中，以儒家伦理学说为主导思想的文化因素占重要地位。其中对民居建筑影响最大的是崇天敬祖思想。

在民居设计中，祠堂、祖堂是建造房屋时首先考虑的内容。古制规定“君子将营宫室，宗庙为先，厩库为次，居室为后”。宋朱熹所著《家礼》一书就有“立祠堂之制”，规定“君子将营宫室，先立祠堂于正寝之东，祠堂制三间或一间”，说明古制对祠堂之重视和限制。

在祀宅合一的民居中，建造时也以祀为主。它在设计时，先将供祀祖先、天地的场所作为祖堂，位置在整个宅第的最后一进的正中厅堂，称为后堂，亦称祖堂。后堂的开间、进深和脊高、檐高都有一定的尺寸规定，甚至神龛、香炉的位置、高度也有所规定，不得随意更改。

还有一种思想影响民居建筑，即风水观念。

风水观，古称堪舆学，它来源于阴阳五行学说，原是古代阳宅和阴宅分别择位定向时考虑气候、地理环境的一门学说。阳宅即民居建筑，例如，在农村，对民宅的选址一般已形成一种比较固定的负阴抱阳模式，即村前要有流水，村后要有高山，房屋坐北朝南，地形前低后高。从现代观念来分析，这种布局原则还是有科学性的一面。譬如村落面靠流水，这是食水、交通、洗濯的需要。村后高山作屏，可抵御寒风侵袭。地形前低后高，说明坡地上盖房子既要求干燥又要易排水，对居住及人体健康有益。

风水观念中还有一种象征和压邪思想，如江南、皖南一带民宅喜用马头墙。所谓马

头墙，就是在山墙墙头部位做成台阶式盖顶，在盖顶之前沿部位，使墙头上翘做成似马头形状，称为马头墙。据当地人讲，山墙作马头形状，说明该户家族中曾有人中举。武官用马头状，称马头墙；文官则用印章、方形，称印石墙。实际上，山墙做成马头墙或印石墙，是显示住户家庭中举、有朝官的一种用建筑表现的炫耀方式。而老百姓家只能用双坡屋面。

广东潮州民居的山墙墙头部分有做成金、水、木、火、土五行也是同样的道理。在实际调查中，民居建筑通常用两种山墙：一种是曲线形，称水墙；另一种是金字形，称金墙。依照五行相生相克学说，水压火是五行相克论说，金生水、水克火，是五行相生又相克的论说。其目的和意图都是为了压火、防火。古代建筑因是木结构营造，最怕火灾，建筑一旦失火，无法可救，但当时科学水平有限，无法采取有效的防火措施，于是采用压邪这种祈望吉祥平安的心理手法，从而可见天地观念对民居建筑的深刻影响。

Text B　中国民居建筑的特征

民居特征，主要是指民居反映出来的本民族本地区最具有本质的和代表性的东西，特别是要反映出与各族人民的生活生产方式、习俗、信仰、审美观念密切相关的特征。

具体来说，民居特征在建筑上主要反映在下列三个方面。

布局特征，主要表现在平面形式丰富和空间组合多变

以汉族民居为例，大型者如多进院落式集居住宅，小型者如三合院或四合院住宅，它们的基本布局都一样，前堂后寝，中轴对称，正厅两房，主次分明，院落相套，规整严谨，外部有高高的封闭围墙、内部则是层层院落，或纵向发展或横向发展，形成一种外封闭内开敞、组合灵活而又紧凑的布局形式。在其他民族中，如云南傣族，在封建领土经济制度下，家庭盛行一夫一妻制，每户要收户头税，年长子女成家后要另立门户，故村落布局中的民居都属独户式竹楼。竹楼内平面是大空间，内部布置以木板相间隔，也只是厅房而已。厅对外，房对内，厅兼作厨房用。这主要是由于小家庭生活方式的缘故。傣族人民虽信佛教，但无家神，寨有缅寺，宅内无供神处。也有见到竹楼内有供神位者，则是受了汉族的影响。

民居的外形特征，主要表现在造型朴实、群体和谐、环境优美以及鲜明的民族特色

我国民居建筑大多选址在优美的山水自然环境之中，以取得舒适的生活条件。农村中建筑群沿自然地势布置，或聚或散，或高或低。但城镇中的建筑则互相毗连，构成街坊。在个体造型上，则根据功能需要，灵活组织空间，无论规则严整还是错落有变，它都顺乎自然，很少矫作，反映出淳朴、真实的面貌。

我国是由五十多个民族组成的国家。各个民族，由于各自经历的历史条件和聚居地区自然条件的差异，也由于民族的生产方式、生活习俗、宗教信仰和审美观念的不同，表现在民居建筑形式上也存在着较大的差别，如藏族的石砌厚墙台阶式平顶建筑、维吾

尔族的内院拱廊式平顶建筑、傣族的独院干栏式竹楼等。

此外，我国地貌山水丘陵地多，地形高差错落，各地气候悬殊，各民族又因生活习俗、文化信仰、审美爱好的不同，因而形成了我国各民族各地区民居的外貌呈现出千姿百态、丰富多彩、灿烂而又鲜明的民族特色。

民居的细部特征，包括装修、装饰、色彩、样式、风格等，主要来源于民族的习俗、喜好、愿望和审美观念。它主要表现为丰富多彩、含蓄得体、艺术和实用的结合并寓有深刻的文化含义

细部中，最突出的部位是大门、门窗、山墙面和某些构件装饰。由于这些构件都位于建筑外表的最显目之处，它最易被人所注目。因而，长期以来就成为民族和地方特征的重要内容。

大门在封建社会下是贫富贵贱等级的一个重要标志。在旧社会，不论贫富家庭都竭尽一切财力，为自己的住宅大门进行装饰和美化，目的是用来显示自己的门第，大门就成为反映民居经济文化的象征。如北京四合院大门、内院垂花门，云南白族民居门楼，广州民居趟栊门等。人们通过大门布置的方式和形象，可以比较容易地识别这些民居究竟是属于哪个民族或哪个地区。

窗户是民居中最常见和常用的一种建筑装修元件。在窗户上，无论是它的大小、式样、色彩，或者是窗棂花纹、工艺，无不反映了人民的喜爱和审美心理，人们从窗户的形式上也可以判断出它是属于哪个民族或是哪个地区。如藏族的密肋饰带窗楣和梯形窗，新疆维吾尔族的长条窗、尖拱窗，北方汉族四合院民居的支摘窗，广府民居的满周窗，中原地区窑洞民居的拱券窗等都是一些比较典型的实例。

色彩、装饰、样式以及某些图案，由于当地民居经常使用，也成为一种独特的艺术表现手段和特征标志。如江南水乡民居喜用灰瓦白粉墙、瓦片编成漏窗图案或各种植物形图案。南方民居喜用青砖墙面、陶塑脊饰。傣族民居喜用编竹装饰，如大象、槟榔树、孔雀或日、月图案装饰等。

此外，在民居中，各族人民常把自己的心愿、信仰和审美观念，把自己最希望、最喜爱的东西，用现实或象征的手法，反映到民居的装饰、花纹、色彩和样式等构件中去。如汉族的鹤、鹿、蝙蝠、喜鹊、梅、竹、百合、灵芝、万字纹、回纹等，云南白族的莲花，傣族的大象、孔雀、槟榔树图案等。这样，就导致各地区各民族的民居更呈现出丰富多彩和百花争艳的民族特色。

Unit 3 The Characteristics of the Art of the Chinese Traditional Vernacular Architectures

（中国民居的建筑艺术特征）

【参考答案】

Text A

Task 1

1. Answer the following questions with the information contained in Text A.

1）The group layouts — harmonious and unified; The image of detached houses — simple and authentic; Space composition — flexible and graceful; Detail decorations — rich and varied

2）Because they show the unique combination of aesthetics and structure adaptive to the conditions of the local climate and geography with local building materials.

3）It represents the characteristics of the people's dignity, imposing boldness and straightforwardness.

4）The characteristics of harmony in the Chinese traditional vernacular architecture are shown as follows: several houses are grouped into one yard and several yards form one homestead; several homesteads constitute a lane, a street or a village, which in turn becomes a part of a town or a city.

Task 2

2. Fill in the blanks with the given words. You may not use any of the words in the bank more than once. Change the form of the given words if necessary.

1）authentic　2）characteristic　3）detach　4）decoration
5）harmonious　6）constitute　7）facilitate　8）layout
9）conspicuous　10）unify

Task 3

4. Translate the following paragraph into Chinese.

河湖地带的民居建筑充分利用水面，或沿河布置，或临靠水面。特别在江南地区，民宅临街背水，建筑与道路、河流走向相适应，创造了方便生活的优美环境，充分反映了江南水乡特色。

Speaking and Writing

1. Please match each letter that shows where each vernacular house is located with the the house in each picture.

A　C　B

【参考译文】

Text A 群体布局及单体建筑形象

中国民居建筑的艺术特征主要反映在群体布局、单体建筑形象、空间组合和细部装饰等四个方面。

群体布局——和谐统一

中国传统民居布局的特点不是单座而建，而是几座合成一院，几院合成一宅，宅合成巷、街、村，再合成镇、城。民居建筑的形象表现不仅是一座建筑所能反映的，而要通过一个院落、一条巷、一条街、一个村落，甚至一个墟镇才能反映出来。

南方平原地区城镇因人口密集，民居建筑大多集中布局，相互毗邻排列整齐，四周街巷围绕，表现出严整、封闭的特点。农村中的民居考虑到便利生产，又要节约农田和方便交通，常在沿河、沿路和坡地建造，建筑有良好的朝向，并表现出一定的规则性。

不少山区或丘陵地带是少数民族集居之地。这里的民居建筑常沿等高线布置，有沿山腰的，也有在山脚的。它们的特点是自由灵活，高低错落，与自然环境协调。在汉族客家山区建造的聚居防御围楼，有单座建造的，也有多座建造的，有圆形的，也有方形的，还有两个甚至多个方形或圆形土楼交错连接的，反映在群体外貌上体形巨大、稳重，气势豪放、粗犷。

河湖地带的民居建筑则充分利用水面，或沿河布置，或临靠水面。特别在江南地区，民宅临街背水，建筑与道路、河流走向相适应，创造了方便生活的优美环境，充分反映了江南水乡特色。

在炎热多雨的粤中地区村落，民居建筑密集排列，像梳子一样规则而整齐，称为梳式布局。它前面常设鱼塘，后面种植竹林，禾埕旁又栽植一两棵大榕树，它与周围稻田结合，反映出村落民居建筑的简朴、自然和一派农家田野风光。

在南方福建、江西、广东和山西、陕西的一些山区，由于防御的需要，民居建筑常集中组合成大型的堡寨或围楼，有单层、二层，也有多层，或圆或方，对外封闭，对内开敞，这类建筑体型巨大、稳重而粗犷。有的土楼几个或更多的成组建造在一起，给人以深刻的印象。

在西南黔桂地区的侗族聚居的村寨，在寨中心多辟出空地修建一座高耸的鼓楼作为公共活动中心，它与村边的风雨桥和整片侗寨民居，构成优美的总体天际轮廓线。

单体建筑形象——淳朴真实

汉族地区单体传统民居一般都是单层建筑，三开间，坡屋顶、白墙灰瓦，在农村居多。也有民居建筑为二层者，大多在城镇人口密集的街巷，其形象都比较淳朴真实。有的民宅在大门、门窗或山墙、墀头部位偶然作一点装饰处理，表示美化。在山区和坡地的一些单体民居建筑，因结合地形，布局比较自由灵活，其外形简朴中带有轻巧。二层或二层以上的坡顶民居，有的每层都设外廊或向外出挑，有的则利用体型组合和挑檐，形式多样，如云贵桂湘各民族地区的吊脚楼就是实例。

四川山区和坡地地带的民居则结合地形，顺坡而建，或建筑顺着地势层层抬起，屋面也做成层层升起的形式，富有韵律感。

在民居单体建筑形象中，最显眼的部位是屋顶。屋顶是利用当地材料、为适应当地气候地理环境条件而将独特的审美和结构相结合运用构造的产物。不同的自然环境塑造了不同的房屋平面，而不同构造采用的不同材料和不同民族生活方式的差别产生了外形各异的屋顶。因而长期来，屋顶形式就已成为各地区各民族建筑的主要特征，如汉族民居中江南的马头墙屋面、广东的镬耳墙屋面、藏族民居的平顶屋、蒙古族圆帐顶、维吾尔族的穹窿顶、傣族的高耸歇山顶及四川民居的穿斗式披梳大屋面等都是非常典型的民居形象特征。

Text B　空间组合及细部装饰

空间组合——灵活优美

民居建筑空间分为外部和内部两部分，外部空间指环境。

民居的艺术魅力有时不是单纯依靠民居建筑本身的表现来达到的，通常还要依靠民居建筑所在的周围环境来表达。

山腰茅屋因挺拔高山才显其清秀、宁静，河畔宅居因潺潺流水才显其潇洒自在，坡地民居因建筑结合地形才显其轻巧、精灵。

汉族民居的内部空间艺术主要表现在两方面：一是庭院空间，二是厅房室内空间。

庭院空间

我国的民居建筑多数以院落或厅堂为中心来组织空间。以北方四合院为例，院内正房坐北朝南，东西厢房沿南北轴线对称布置。南面设游廊、花墙和二门，二门外是东西狭长的前院，前院南面是倒座房，大门设在东南角。这种空间组合方式，主次分明，分区明确，既能满足长幼有序、内外有别的使用要求，同时，也为内院创造了安静的居住环境。

南方汉族民居庭院较小，称为天井。也有稍大的庭院，通常作为庭园。小型天井或庭院一般不栽树，因树大、杆粗、叶密，遮挡阳光且不通风。有时栽种一两株单株细木或栽竹，也有置盆景者，总的来说以绿化为主。

中型庭院可植树一株，并辅以假山或绿化，它与建筑檐廊相连接，形成宁静、安逸的气氛。

大型庭院可置假山、池水、花木，再建楼阁，或亭台，或舫榭，它与宅居相连，组成一组比较完整的住宅园林。私家园林中，引进大自然山水景色，划分大小不同的景区，再运用对景、借景手法，这样就可以在有限的空间内，获得可居、可游、可行、可望的艺术享受。

厅堂内部空间

一是门窗、隔扇，这是厅堂直接面向庭院天井的部分。隔扇的花纹图案丰富多变，是民居中是最富艺术表现力的构成之一。

在大型民居建筑中，门窗隔扇的格芯常用木条拼成方格纹、藤纹和锦纹，也有雕人物花鸟者。有的在槛窗下半部另加罩格，作遮挡视线用。它通过大面积的图案、纹样和通透光影的对比来取得装饰装修艺术效果。在小型民居建筑中，往往采用窗下突出宽窗台，窗上加楣檐，如临水则突出窗栅，或在楼层挑出栏杆等手法，来增强建筑外观上的凹凸变化和虚实对比，既符合使用要求，又增加艺术气氛。

二是在厅堂和房间内部为了分隔空间，通常采用屏、罩、隔断等构件，其雕刻技术与精美图案，很有特色。

三是厅堂或廊檐的梁架及其附属构件如柁墩、梁头等，在不影响其结构性能要求下施以雕琢，丰富了空间艺术效果，也增加了美化作用。

四是匾额、楹联，这是厅堂空间中不可缺少的内容。我国古代中原的世家望族，南迁后仍然保持着家族的历史流传，例如把堂号名称写在匾额上并悬挂在厅堂正中或门楣上。此外，也有在厅堂明间金柱上挂上对联，称为楹联。此外，在园林中的馆、轩、楼、阁内也大多设置有匾额、楹联，不但增加了园林的文化气息和诗意，同时，也丰富了园林空间的艺术特色。

五是家具、陈设，它是厅堂和住房中必不可少的用具，也是丰富室内空间的重要手段。如厅堂中，为接待宾客而设置了桌、椅、案、几；为祭祖祀神而设置了神案、神龛；夜间为了照明而设置了吊灯、灯笼。在住房中则设置了供生活使用的木床、木橱、木柜和桌椅。这些用具，不但具有实用价值，而且雕作精致，富有艺术和民族特色。

细部装饰——丰富多变

细部处理

细部处理在民居建筑的艺术表现中占有重要的地位。其目的主要是为了强化建筑整体或局部的艺术效果。它一般都在实用部位上采用，特别是人眼可以直接看到和人手可以直接摸到的地方，更是细部处理的重点。民居建筑的表现部位一般都在大门、墙面、地面、梁架、柱枋、楼梯、柱础、栏杆、台阶等处。

民居建筑的大门，历来是户主显示其社会与经济地位的标志，因而，许多地区都对大门的式样、用料、工艺、装饰、色彩精心经营，以达到突出门第的作用。

北京地区民宅大门常用一座单独门屋建筑，正中双扇黑色大门，门上安置铺首铁环，虽然装饰不多，但它外貌严肃封闭，使人感到阴沉可怕。

江南地区一些富裕大户或文人、士大夫宅第常用牌楼作为大门，门楣上加以题词，门檐做成挑檐式，并用青砖砌筑和精致刻砖雕饰，以显示其高贵与文雅，当地称为门楼。皖南地区古老村落，有的用牌坊作为进村的大门，如安徽歙县棠樾古村，入村前先看见牌坊就是一例。有的地区一些民居大门做成凹入形，称凹斗门，既防雨，又避晒。粤中地区城镇民居则采用一种名叫“趟栊”形式的木制门，它是在黑漆双扇大门外，加上用木横条组成的一种栅门。当大门敞开时，趟栊门关闭，目的是通风，又防盗。有的在外面还装上四扇半截式木制通风栅门，更可达到通风和遮挡视线的效果。这种栅门木质坚固，图案优美，寓艺术与实用于一体。

云南白族民居八字门楼是它最突出的标志之一。门楼平面外开八字形，墙体用砖贴面。屋面有主次，高低有错落，墙楣、墙身、屋檐装饰艳丽。

傣族竹楼大门敞开式置于竹楼二层，它由木梯直上，经廊道，即见大门。大门由竹编制而成，有的还编制图案，简洁又美观。

民居建筑中，除大门外，还有二门，这是封建制度下区别内外的标志，在富裕大户中对此二门要求比较讲究和严格，如北京宅第中的垂花门。此外，还有一种洞门，多数在园林中采用之，其手法比较轻巧、自由，外形可做成月形、圆形、瓶形、壶形等。

墙面处理也是民居建筑细部艺术处理的重要手法之一，它是依靠所用材料本身的质感来取得对比效果的。例如江南和川东地区用栗色木柱木骨架和划分白色粉刷墙面以取得素雅效果。在闽南泉州一带，则用石块和红砖穿插砌筑，或在红砖墙上镶嵌石框和用深浅色砖砌成图案以取得质感对比的艺术效果。广东潮州地区民居建筑，在墙尖下垂带部位做成层层线条，以变化的轮廓线来取得艺术效果。福建泉州杨阿苗宅中的小院墙面上，有圆形砖雕一幅，在壁面上部又有砖雕楣檐一幅，图案优美，雕技精湛。

民居建筑中，院落地面和檐廊地面，常用砖、瓦、石材铺砌，有的还在石材上施以雕刻，有的用卵石拼砌出各种图案，起到美化作用。特别在庭园、斋轩等建筑中较多采用。

柱础有承重和防潮、防水作用，也是装饰部位之一。它常做成鼓形、瓜形、筒形、瓶形、斗形、八角形等。有做成单层的，也有做成双层的。在柱础的表面上通常进行雕刻，有雕花卉鸟兽者，也有雕几何图案者，其工艺和题材，都有明显的地方色彩。

栏杆有木制、石造两种。木制栏杆较多用于室内，如廊下、楼梯、二楼柱廊，或楼层周围悬空部位，如楼井等，可避日晒雨淋。其处理原则是实用为主，并与艺术相结合。栏杆不仅实用，其艺术处理也不违背木材的结构性能，崇尚朴实并以线条为主，适当加以浮雕或浅雕装饰。石造栏杆主要用于室外临水部位，由于水面空间不会太大，如跨水之桥，其体量也不会太大。因此，石造栏杆宜矮、宜小和坚实稳重，甚至有的园林中临水栏杆可用条石做成，人过桥时还可以在石造栏杆上休息，遥望池水彼岸景色，也是一种艺术享受。

台阶、台基位于地面上，有防水防潮作用。它虽在低处，但当人们走过时，常怕摔倒而特别注意。因此，在台阶、台基上一般不作细部处理，还它自然面貌。如需处理，也是简化，略作线条加工而已。

装饰装修

装饰是附加在构件上的一种艺术处理，如屋面上的脊饰、大门装饰、外檐装饰、山墙墙面装饰、室内梁架装饰等，它们有实用价值，也不影响建筑物的使用和结构，目的是为了美化建筑物，在封建社会中它还是显示户主地位和财富的标志。

装修也称小木作，主要指室内布置和陈设，它包括门窗、隔断、屏罩和家具等，兼具实用价值和欣赏价值。

装饰是建筑艺术表现的重要手段之一，其特征在于充分利用材料的质感和工艺特点来进行艺术加工，以达到建筑性格和美感的协调和统一。

装饰的工艺特征则是充分运用刀、锤、斧、锯等工具，直接在材料上进行构图和艺术加工，根据不同的材料采取不同的方式，从而形成不同门类装饰的艺术表现和风格。

装饰还有一个明显的意向特征。它将我国传统的象征、寓意和祈望手法与民族的哲理、伦理思想和审美意识结合起来实现艺术表达。这种象征手法在民居装饰中较多采用，通常是用形声或形意的方式来表现。形声是利用谐音，通过借某些实物形象来获得象征效果。如用莲、鱼表示连年有余，蝙蝠、梅花鹿、仙桃表示福、禄、寿等。形意则是利用直观的形象表示本身意义的内容，如松鹤表示长寿、牡丹表示高贵、莲花表示洁净、梅竹表示清高亮节。也有将形声和形意的内容合在一起使用的，如在瓶上加如意头寓为平安如意。这些图案花纹大多反映了人们的吉祥愿望，是一种具有民族特色的文化传统和观念的体现。

在民居中，装饰手法非常丰富，一般来说，它贯彻了三个原则：一是实用与艺术相结合；二是结构与审美相结合；三是综合运用其他艺术品类如绘画、雕刻、书法以及匾额、楹联等民族文化艺术的特长。这样，就增加了装饰艺术的民族特色和它的特殊感染力。

此外，装饰装修艺术表现手法上还有下列特点，即构图形象上的丰富和统一，题材内容上的多样化，既可用历史人物，也可用动物、植物和花草，不拘一格。在色彩处理上，以典雅朴实为主，重点部位则稍加突出。

民居装饰在部位安排上分为室外和室内，室外以大门、屋脊、山墙和照壁为主，室内包括门窗、隔扇、梁架等。以工艺类别来说，分为雕和塑两大类，包括木雕、砖雕、石雕、灰塑、陶塑、泥塑，粤东沿海一带还喜欢用嵌瓷装饰，它对防海风侵袭很有效果。

综合上述南方汉族民居建筑的主要内容和艺术表现手法来看，它的总特征可以归纳为下列几点：

1）布局上的规整性、类型上的丰富性和组合上的灵活性。

2）民居在适应气候、地理、地貌、材料、结构等自然条件下因地制宜、就地取材、因材致用的做法是非常突出的。

这是因为，民居是由人民建造的，设计者、施工者、使用者都是人民自己，三位一体。中国封建社会下人民大部分是农民和城市庶民，户主的经济能力和民居的使用要求决定民居建筑一定要走勤俭节约的道路。

3）外形朴实、群体和谐、装饰装修丰富多彩。

中国封建社会在“藏而不露”思想支配下，民居外形是朴实的。民居的艺术表现难于在单体建筑中表达，而只有在群体中才能得到体现。古代“中和”思想在民居建筑中也深受影响，例如平面布局中的中轴对称，地形处理中的前低后高、前水后山、左右环抱，在形象造型中的大小对比、稳定平衡等，都明显地表现出和谐的特点。由于民居使用者的经济和地位的不同、文化素质的差异以及居住在城乡地域不同等因素，建筑的等级制度除了规模、体型、开间等标志外，很大程度是用装饰装修和细部来表达的。我国匠人高超熟练的工艺技巧水平给建筑装饰装修表现提供了可能。民居装饰装修的品类齐全，构思独特，题材广泛，手法多样，它为民居建筑艺术表现增加了无限的光辉和特色。

Unit 4 Tianjing Courtyard Houses（Ⅰ）
（天井民居 Ⅰ）

【参考答案】

Text A

Task 1

1. Answer the following questions with the information contained in Text A.

1）First, the courtyards are small. Second, there are many lanes and alleys. Third, open halls and concealed chambers are the basic elements in the layout of the traditional vernacular architecture.

2）The courtyard in South China takes up little land, and its layout is compact, with buildings adjacent with one another.

3）There are not only lanes and alleys that function as the connection among the buildings within a homestead, but also lanes and alleys that are called fire lanes, or passageways to connect homesteads with the function of fire prevention and ventilation.

4）The partition board will be removed so that the hall and the small courtyard can be combined as one bigger space to accommodate more clansmen.

5）The vernacular architecture in South China generally has very good orientations so as to obtain a better condition of facing the wind. At the same time, a very good ventilation system is formed through an open hall, a gallery or corridor and a tianjing courtyard.

Task 2

2. Fill in the blanks with the given words. You may not use any of the words in the bank more than once. Change the form of the given words if necessary.

1）adjacent　2）exposure　3）accommodate　4）dampness　5）compact
6）scorching　7）indispensable　8）layout　9）conceal　10）drainage

Task 3

4. Translate the following paragraph into Chinese.

南方汉族民居与北方一样，平面布局都是院落式。但是，南方地区由于气候湿热多雨，地理环境多丘陵和河流，加上人口稠密，土地资源紧张，因而，民宅占地少，布局也较紧凑，房屋毗连，称为天井式民居。

【参考译文】

Text A 南方民居建筑

南方汉族民居与北方一样，平面布局都是院落式。但是，南方地区由于气候湿热多雨，地理环境多丘陵和河流，加上人口稠密，土地资源紧张，因而，民宅占地少，布局也较紧凑，房屋毗连，称为天井式民居。

南方民居有下列特征：

一是小天井。南方天气炎热，民居要求有阳光，但又怕日晒太多，小天井可满足这要求。此外，天井又是通风、换气、采光、排水的场所，还是民宅内交通会合的地方，是传统民居中不可缺少的元素。

二是多巷道。民居建筑群中的联系主要靠通道，称为巷道。巷道有几种：露天的称巷，也有的称弄，有盖顶者称廊；可见到天空的又能避雨淋日晒的称檐廊或敞廊，见不到天空的称内廊。廊巷不仅作为民宅室内房屋之间的联结，在宅与宅之间的联结也有巷道，称为火巷，或称夹弄，它还兼有防火通风之用。

三是敞厅暗房，这是传统民居布局中的基本元素。厅堂是民宅中最重要的和必不可少的公共活动场所，是家族文化的核心和表现，一般都位于中心部位，面向天井，厅前的格扇采取活动敞开式。家庭有重大事情，则把格扇拆下，使厅与天井合一，成为大空间，可以容纳更多的族人。

民居建筑组合时，可纵向，可横向，也可纵横结合，类型丰富，组合灵活。天井庭院还进行绿化，外墙封闭，宅内开敞，室内外空间紧密结合。

南方民居一般都选择良好的朝向，以便获得更好的迎风条件。同时它通过敞厅、廊巷和天井，组成一个良好的通风系统。实践证明这是南方天井式民居持久延续的主要原因之一。

天井式民居的艺术表现，主要靠建筑群体艺术、空间处理、材料质感以及室内外装饰装修和家具陈设布置。

Text B 江苏苏州东山雕花大楼

苏州东山雕花大楼建成于1924年。因建筑主楼的梁桁、门窗及门楼雕刻精美华丽，人们俗称雕花大楼。整个建筑集砖雕、木雕、彩绘、泥塑诸艺，巧夺天工，精妙绝伦，为我国雕刻艺术中的一件精品。

雕花大楼坐西朝东，平面呈多边形四合院式，两进五开间，内有门楼、主楼、花园。大门外建有一道“之”字形的照壁，上嵌砖刻“鸿禧”，雕琢精致。

入口门楼正面为单坡瓦顶花岗石门，水磨青砖拼贴镶嵌。门楼的上、中、下枋，均饰有画像砖雕，其浮雕、圆雕、透雕和谐相融。前楼大厅是大楼雕刻最多的地方，其木雕雕饰丰富，技艺精湛。

整座雕花大楼充分展现出江南传统民间雕刻艺术精华，它集建筑、雕刻、书画、文学、工艺、园艺于一体，因而被誉为“江南第一楼”。

Unit 5 Tianjing Courtyard Houses（Ⅱ）
（天井民居 Ⅱ）

【参考答案】

Text A

Task 1

1. Answer the following questions with the information contained in Text A.

1) The Suyongtang Hall is the public hall for the big clan of the Lu's, built in the Ming Dynasty, with a very prominent scale and social status. It is as wide as three bays, with high eaves and a high ridge. The main hall is connected with the back hall by a hallway, forming a 工-shape plane.
2) The village is surrounded with 8 hills and its layout takes the Zhong Pond as the center and the houses are of the radial form. The 8 lanes spreading outward divide the whole village into 8 sections so as to form the pattern of interior Eight-Diagram battle formation. Zhong Pond in the center of the village, a big pond of Eight-Diagram battle formation, looks like a diagram of supreme pole with Yin and Yang, the half with water as Yin and the other dry half as Yang. The water in the pond adds mystical coloration and vivid anima to the village.
3) Lishui Street of Yantou is located on the west bank of the middle reaches of Nanxi River, and was built in the early Tang Dynasty. The place was the only passage for salt merchants.
4) It is a chair along Lishui Street, winning its name for being seated by beauties and leaned along their waist.

2. Fill in the blanks with the given words. You may not use any of the words in the bank more than once. Change the form of the given words if necessary.

1) exquisite　2) surrounded　3) located　4) reign
5) diversify　6) reflect　7) consist　8) solemn

Task 2

3. The following English sentences on the left show some different ways of clarifying

locations. Read the English sentences carefully and translate the Chinese sentences on the right into English by simulating the structure of the English sentences.

1) Located in the middle reaches of Nanxi River, Yantou Town is named after its location at the foot of Furong Mountain.
2) Nanping Village, located four kilometers southwest of Yixian County, is named after the mountain it lies against in the southwest.
3) Facing the main hall stands the arch gateway with meticulous brick carvings on it.
4) At the entrance of each building, there is a stone plaque in the upper side of the door.
5) The north of the village is an ancient ferry terminal, called "Huai Xi Shou Ji".

4. The following English sentences show some different ways of clarifying the construction time of a building. Read the English sentences carefully and translate the Chinese sentences by simulating the structure of the English sentences.

1) Zhonglou Village, located in Taiping Town, Conghua City, was built by the Ouyang clan during Emperor Xianfeng's reign of the Qing Dynasty.
2) At the northeastern end of the pool, the "Shui Yue Tang" was built in the reign of Emperor Huizong in the Song Dynasty (1119 – 1125).
3) Chengzhi hall, built in the 5th year of Emperor Xianfeng of the Qing Dynasty (1853), is a three-hall and two-story wooden structure in three-room breadth.

Task 3

5. Translate the following paragraph into English.

The Complex of Zizheng Official Hall, located in the west of Sanhua Village, Xinhua Town, Huadu District, Guangzhou City, was built during the second and third year of Tongzhi reign of the Qing Dynasty (1863 – 1864). It has a history of more than 150 years.

【参考译文】

Text A 浙江省部分天井民居

浙江东阳卢宅

卢宅位于东阳市吴宁镇东门外，这里丘陵起伏，河道纵横，面山环水，环境幽雅。

卢宅创建于明永乐年间，现存建筑乃明清两代相继兴建。卢宅是一处由数条南北向纵轴线组成的建筑群，房屋数千间，占地15万平方米。主要建筑是街北的肃雍堂建筑群。

进入卢宅大夫第建筑门楼，只见双扇黑漆大门，双层砖挑屋檐，旁为灰白墙面衬托，形象朴实。

肃雍堂是卢姓大族的公共厅堂，明代所建，其规模和地位十分突出。肃雍堂面阔三间，檐高脊高，廊庑式建筑。大厅与后堂用穿堂相连，形成工字形平面。肃雍堂共九进院落，是卢宅保存最完整的一条轴线。屋内梁架、柱枋、挑檐等木雕装饰工艺精湛，题材丰富、构图和谐，显示了浓郁的地方特色文化。

浙江兰溪诸葛村

诸葛村位于浙江兰溪市，古称“高隆”，汉代诸葛亮后裔在此定居后，即按照先祖创造的九宫八卦阵式，规划和建造村庄。

诸葛村自然环境好，村后高山，村前平原，可耕可樵，可渔可猎，且交通方便，适宜人居。村落位于8座小山的环抱中，其布局以钟池为中心，房屋呈向心状，向外延伸的8条巷道，将全村分为8块，从而形成了内八卦布局模式。村中心的钟池为八卦式的大池塘，形似太极阴阳图形，有水的一半为阴，另一半干地为阳，池水给诸葛村增添了神秘的色彩和活泼之灵气。

诸葛村现存有明、清宅居建筑200多座，布局奇巧、结构精致。丞相祠堂建于明万历年间，建筑中庭选用了4根直径约为50厘米的松木、柏木、桐木和椿木，寓意“松柏同春”。整个祠堂翼角高翘、雕刻精美、造型庄重、气氛森严。

浙江永嘉岩头丽水街

岩头丽水街位于楠溪江中游西侧，始建于唐初。塔河庙由庙、河、戏台、接官亭、古树等组成，是村民大型的户外活动中心，堪称风景园林。

东园的蓄水堤上，建有一条古商业街——丽水街。300多米长，90多座相连的一、二层商店，对水一字排开，临水有美人靠。历史上为盐商必经之路，兼有小憩和商业作用，也是一处极具特色的景观街道。

街南端是寨墙的南门，门边高阶上有凉亭，离亭50米处有一座厚重又灵巧的花亭（又名接官亭），四方二层，底层木板墙，二层外挑，通体花窗，重檐攒尖顶，朴素庄重，与丽水街相互呼应。

Text B 其他南方省份的典型天井民居

安徽黟县南屏村

南屏村位于黟县县城西南4公里处，因村西南背倚南屏山而得名。

南屏村高墙深巷，全村1 000多人，却有36眼井，72条巷，300多幢明清古民居。全村巷弄，长短不一，弄弄相通，拐弯抹角，纵横交错，有“江南迷宫”之称。

南屏村至今乃保存有相当规模的宗祠、支祠和家祠，被誉为中国古祠堂建筑博物馆。叶氏宗祠“叙秩堂”位于南屏村中心，始建于明成化年间，坐东朝西。大门两侧有一对一人多高、用黟县青石精雕细刻的巨大石鼓，乌亮厚实，非常威严。从大门口可以一直望见上厅享堂的香火案壁，显得轩敞又幽深，使人顿生肃穆、神秘之感。

位于村庄上首的半春园，建于清光绪年间，是村中一所私塾庭院。园中种植了大量名贵花木，尤以梅花为最，故别名为梅园。

江西婺源李坑村

“坑”在赣中就是溪的意思，婺源所有的村庄都建在碧波荡漾、纵横曲折的溪流边上。秋口镇李坑村坐落在一条狭长的山坞之中，建于北宋年间，村落群山环抱，山清水秀，风光旖旎。村内260多户大都沿河而居，每户门前都有一座青石板桥，真是名副其实的小桥流水人家。

李坑村民居属于徽派建筑，除了具有徽派建筑的典型粉墙、飞檐、翘角外，木、石、砖三雕也可称三绝。村内明清古建遍布、民居宅院沿苍漳依山而立，粉墙黛瓦、参差错落。村内街巷溪水贯通、九曲十弯。青石板道纵横交错，石、木、砖各种溪桥数十座沟通两岸，构筑了一幅小桥、流水、人家的美丽画卷，是婺源古村落中的一颗灿烂明珠。

福建连城培田村

培田村位于闽西山区连城县宣和乡河源溪上游的吴家坊，已有800多年的历史，是一处保存较为完整的明清时期客家古民居建筑群。村落由学堂书院、宗庙牌坊、民居祠堂等建筑和一条千米古街、五条巷道、两条贯穿村落的水圳组成，规模宏大，布局讲究，配置得体，壮观和谐。

培田古居民群以“大夫第”“衍庆堂”“官厅”等为代表，是著名的“九厅十八井”式客家建筑。“大夫第”又称“继述堂”，建于1829年，历时11年建成。“衍庆堂”为明代建筑，建筑结构与“大夫第”大体相同，但门外荷塘曲径，门前石狮威镇。“官厅”因接待过往官员而称“官厅”，高墙耸立，四周封闭，设计精巧，工艺精湛，其左右花厅则专供主人休闲会友，楼下厅为学馆，楼上厅为藏书阁。民居布局尽管厅多井多房多，却井然有序，厅与厅之间既有通道相连，又有门户隔阻，各成单元，既利于大家族聚族而居，又不妨碍小家庭各享天伦。

湖南凤凰古城民居

湘西凤凰古城因其西南方有山如凤形而得名。自古以来这里就住着土家族和苗族的先民，至今已有三千年的历史。

凤凰古城坐落在沱江河畔，碧绿的江水从古老的城墙下蜿蜒而过，河畔上的吊脚楼房轻烟袅袅，景色轻盈秀丽极富节奏感。现今的红色砂岩城墙乃清初康熙年代所始建，古城依山傍水，城内民居纵横毗列，极有规律。

古城民居布局大体沿袭传统院落式建筑，如沈从文故居是典型的南方四合院建筑，分前后两进，穿斗式木结构，马头墙上装饰有鳌头，故居小巧别致，门窗镂花，古色古香，清静典雅。

在陈氏老宅典型院落式民居中，天井周围下有回廊，上为跑马廊。回廊右侧有木梯上楼。宅院雕花门窗甚为精美，室内装修工艺精细。

Unit 6 Chinese Traditional Courtyard Houses（合院民居）

【参考答案】

Text A

Task 1

1. Answer the following questions with the information contained in Text A.

1）The spirit of rites and ethics.

2）The central kernel courtyard is in the north-south orientation.

3）The main block is the principal building, comprising the living room and bedrooms of the owner and his wife.

4）The wing-blocks are located on both sides of the main block, generally serving as the bedrooms of the owner's children.

5）The characteristics of the courtyard houses in north China are front-hall-rear-room, the central axis being symmetric, the plane being regular and tidy, the exterior being prudent and enclosed, reflecting a profound hierarchical perception.

Task 2

2. Fill in the blanks with the given words. You may not use any of the words in the bank more than once. Change the form of the given words if necessary.

1）occupying　2）characterize　3）classify　4）axis　5）homestead
6）reflect　7）layout　8）corridor　9）status

Task 3

4. Translate the following paragraph into Chinese.

北方合院式民居的特征是，前堂后寝，中轴对称，平面规整，布局严谨，内部院落相间，外观稳重封闭，反映出深刻的等级观念。

Task 4

5. The following English sentences on the left show some different ways of clarifying layouts. Read the English sentences carefully and translate the Chinese sentences on the right into English by simulating the structure of the English sentences.

1）The Mou-clan Manor is located in Du Village, an ancient town in the north of Qixia County. The Manor, facing south, 158 meters long from east to west, 148 meters wide from north to south, is divided into three groups and six courtyards, with more than 480 halls and main rooms.

2）The Residence of the Wang Family is located in Jingsheng Village, a total of 123 large-and-small-sized courtyards and 1,118 rooms, covering a total area of 250,000 square meters.

3) Ma Architectural Complex is divided into south region, central region and north region, consisting of 6 groups and 22 yards, and covering an area of 20,000 square meters with a floor area of 5,000 square meters.

【参考译文】

Text A　合院民居简介

汉族民居中，宗法礼制、伦理思想占主导地位。反映在平面布局中，多进院落式平面为其主要模式。由于我国地域辽阔，气候差异大，故平地民居可分为北方和南方两大类，北方称合院式民居，南方因人多地少，气候又湿热多雨，房屋毗连，院落较小，故称天井式民居。

北方城镇合院式民居由正房、厢房、耳房、倒座房、后罩房、大门、垂花门、抄手游廊、窝角廊、穿廊、影壁、院墙等单体建筑和建筑构件所组成，其组合形式有单进院、二进院、三进院或多进院。一般宅居是沿着中轴线布置成为纵向一路，大型宅第在主轴旁带跨院，也可二路并列，也有因地形关系呈现多路交错组合的。

四合院的平面布局，正中核心庭院为南北向，采取一正两厢加垂花门或过厅，成为近正方形的格局。正房、厢房、倒座等各房，各不相连，分布在院落周围。四合院平面组合中轴对称，院庭宽敞，阳光充足。正房是合院主体建筑，为主人卧室及生活起居用。厢房位于正房的两侧，一般为儿女用房。

四合院临街一面，布置倒座房，其前檐朝内院，后檐对胡同（或街巷）。宅大门即位于倒座的东面。倒座不开窗，或开高窗，外貌呈现高度封闭感。

最后住宅一进为后罩房，作杂务院用，如临街巷，可辟后门。

北方合院式民居的特征是，前堂后室，中轴对称，平面规整，布局严谨，内部院落相间，外观稳重封闭，反映出深刻的等级观念。

Text B　若干典型合院民居

北京四合院

北京四合院除具有北方合院式民居的前堂后室、中轴对称、院落相套特征外，其建筑艺术形象主要表现在厅堂、大门、二门、宅内的空间组织，庭院绿化以及室内的装饰装修上。

大门是整个四合院的起点，也是整座建筑外向形象和艺术表现的重点。因而，大门的形制、规格成为全组建筑的等级表征，又是社会地位的门第标志。宅第主人为了安全与防卫需要，除加大大门的规模外，还增加了门饰、铺首、金环、门簪等附加物，门外还辅以石鼓、石兽，甚至雕花门枕石。

位于主人内院之前的出入门口，称为垂花门，也是大宅的二门，一般比较华丽，是四合院内的装饰重点。

四合院院落空间组织充分运用廊、墙，或间隔，或相连，形成穿廊、花墙，院内则栽植少量树木，或置盆景，呈现出幽静、安逸、舒适的气氛。

天津石家大院

石家大院位于天津杨柳青镇，是清代津门八大家之一石万程第四子石元士的住宅，堂名“尊美堂”。该宅始建于1875年，南北长96米，东西宽62米，占地6 080平方米，建筑面积2 945平方米，是天津典型的四合院住宅建筑。

宅居的总体布局由东院、西院和跨院组成。

东院为住宅院落，由前后三套四合院组成，各院正房五开间，中间的明间为穿堂。东、西厢房均三开间。西院是厅堂院落，由两座回廊院连接而成，是石家大院中规模最宏大、装修最精美的一组建筑。

南面由两座青瓦硬山顶和卷棚顶前廊构成，前面为花厅，作为接待贵宾及宴会时用，后面为戏楼。

内蒙古呼和浩特恪靖公主府

固伦恪靖公主府位于内蒙古自治区呼和浩特市，古归化城城北。固伦恪靖公主下嫁之际，正是喀尔喀蒙古时局动荡之时，而归化城局势稳定、商业繁荣，是大漠南北重要的物资集散地，加之距京城较近，成为建府首选。

公主府始建于康熙三十六年（1697年），占地1.8公顷，建筑面积4 800平方米，房舍共计69间，其规模超过当时的归化城都统衙门。

公主府由皇家督造，依朝廷工部大式营建，以大青山为屏，札达海河与艾不盖河环抱。其风格与明末清初京畿地区王府相似，采用中国古代建筑体系中传统的中轴对称建筑格局，大面积夯筑地基，硬山式建筑特征。

府第为四进五重院落，前有影壁御道，后有花园、马场、府门、仪门、静宜堂、寝宫、配房、厢房、后罩房依例分布，是目前保留最完整的清代公主府第，也是最为典型的清代四合院建筑组群。

山东曲阜孔府

阙里圣人家——孔府位于山东曲阜东华门大街。孔府坐北朝南，横向分东、西、中三路布置，共有房屋400余间，占地240多亩。中路是孔府的主体部分，前后九进院落，前为官署，后为内宅。

孔府大门三启六扇，二门也是三启六扇。二门院内，柏桧参天、灌木繁茂。过二门后，经重光门，到大堂。大堂后为二堂，有通廊相连，呈工字形平面。堂之左右为厅。三堂六厅之后便是内宅，充分体现中国古代前堂后寝制的布局模式。

内院分前上房、前堂楼、后堂楼、佛堂楼等，为家眷住房。堂楼为两层，楼上有回廊、栏杆。内宅建筑简洁大方，浑厚实用。中路最后是后花园。

山西祁县乔家大院

乔家大院始建于清乾隆年间，是一座城堡式建筑。大院四周是全封闭的高墙，上层是女儿墙地垛口，还有更楼、眺阁，很有气势。宅第大门坐西向东，大门对面是砖雕百寿图照壁。大门内是一条石铺地甬道，甬道尽头是祠堂。

整个宅院分为6个大院，分置两边，北面3个大院，布局是本地典型的“里五外三穿心楼院”，即里院正房、厢房各有5间，外院厢房却是3间，里外院之间有穿心厅相连。南面的3个大院都是二进四合院。6个大院，各由三五个小院组成，院中有院，院中套院。各院房顶有走道相通，便于夜间巡更护院。

乔家大院设计精巧，建筑考究，规范而富有变化，既有整体美感，局部又各具特色，无论是砖雕、木雕、彩绘，还是建筑上的斗拱、飞檐、门窗等均形式各异，变化多端。

陕西韩城党家村民居

党家村始建于元至顺二年（1331年），此后有三次大规模建设。

村坐落在黄土塬间沟谷中，村南有泌水绕行。依塬傍水，既可屏障寒风袭击，获取良好日照，又依据地势组织排水。虽处黄土地区，但风尘较少。

党家村风景优美，瓦屋千宇，绿树成荫，村落、塔碑、楼祠与民居辉映。寨墙和险峻崖身高出村址30多米，极为壮观。村中昂首挺立的望楼、塔碑，构成村落优美的天际轮廓线。

党家村巷道路面石墁，巷道丁字形，巷不对巷，门不对门，而且大门不冲巷口。进村的巷口有防盗哨门25处，极有利于防御。

党家村宅院的门楼高大，门楣上有精美的木雕和显示地位的匾额，门两侧的墀头和屋脊上都有精致的砖雕装饰。

河南安阳马氏庄园

马氏庄园位于河南安阳市蒋村镇西蒋村，为清末兵部侍郎、都察院右副都御史、广西、广东巡抚马丕瑶的大型府第。建筑群分南区、中区、北区，共6组，22个院落。建筑面积5 000平方米，占地面积20 000平方米。门、厢房、堂、廊、楼计308间，规模宏伟，系中原最大的清代官僚豪宅。

庄园保存基本完整。北区为其祖居地，建筑坐南朝北，是一座较大的两进四合院；中区建于清光绪年间，规模最大，东西并排四路院落组群，即沿门前东西大道并列四路院落，均坐北朝南。当地民众习惯把传统建筑纵向中轴线统领的宅院称为“路”。西边三路院落以中路为主院，两边为次院，属一主两次并列的深宅大院。这三个院落的格局相近，全部由四进院落组成，东、西院面阔略小于主院。最东面的二进四合院为马氏家庙和家族学堂；南区建于民国初期，其格局、风格也与中区建筑群相近。

Unit 7 Crisscrossed-wood Structure（干阑民居）

【参考答案】

Text A

Task 1

1. Answer the following questions with the information contained in Text A.

1）On the boundary areas of Guizhou, Guangxi and Hunan Provinces.

2）The relief of the land is high and precipitous, the forests dense and thick, with a crisscross network of rivers and streams, and a damp and hot climate.

3）Dong people.

4）With simple and natural decorations, but without paintwork, an overhanging house presents lissome, vivid, flexible and unaffected style and features.

Task 2

2. Fill in the blanks with the given words. You may not use any of the words in the bank more than once. Change the form of the given words if necessary.

1）inhabitants	2）teeming	3）vernacular	4）lissome	5）stretched
6）situated	7）dwell	8）precipitous	9）prop up	10）detached

Task 3

4. Translate the following paragraph into Chinese.

这种楼被称为吊脚楼。一般有 2 ～ 3 层，木构架、木楼板、木梯、干阑式构造，底层养牲畜，楼上住人。吊脚楼前有村道与公路相通，其后建于山坡，有的山坡较陡，当地村民就用细长的木柱在山坡上支撑着大屋檐下的大木楼，呈现出既惊险又稳定的粗犷、淳朴外貌。

【参考译文】

Text A 干阑民居简介

我国很多少数民族居住在西南山区，不少地方是山坡地带，当地盛产木、石，因此，他们创造了依山坡地带的半坐半悬、用木柱支撑悬挑的住屋，一般有 2 ～ 3 层，木构架、木楼板、木梯、干阑式构造，底层养牲畜，楼上住人，称为吊脚楼。吊脚楼前有村道与公路相通，其后建于山坡，有的山坡较陡，当地村民就用细长的木柱在山坡上支撑着大屋檐下的大木楼，呈现出既惊险又稳定的粗犷、淳朴外貌。其中侗族、苗族、壮族等民居建筑中的吊脚楼就是典型的实例。

侗、苗、壮族主要分布在贵州、广西、湖南交界的地区。地势高峻，森林茂密，溪流纵横，气候温湿。他们聚族而居，有大寨、小寨，其布局无一定规则，大多选择近山傍山之处，建筑随地形自由伸展。住户为单栋木楼、多层，户户紧密连在一起，中间只留狭长小道。但寨中必有一大块作为公共活动场所的空地。侗族寨内在这块大寨地上就建有一座高大多层立柱式建筑，称为鼓楼，这是同一村寨、同一族化的社会、政治、文化方面聚众议事的活动中心，又是文化娱乐中心，它是侗族聚落的重要标志，代表着侗族文化，称为“鼓楼文化”。

风雨桥是横跨溪上的交通建筑，既是跨河通道，又是族内居民平时休息交往的空间，节庆时也是唱歌饮酒、吹奏芦笙的地方，它位于入口处，是进寨的标志。

侗、苗、壮族民居是三层穿斗式结构干阑式木楼，当地称为吊脚楼。其上两层通常从架空层挑出，或逐层向外挑出，设外廊，屋顶为悬山，盖瓦或覆杉皮，坡度平缓，出檐深远。吊脚楼外形上大下小，上实下虚，装饰简朴自然，不施油饰，呈现出轻盈、生动、灵活、质朴的风貌。

Text B　湘西土家族民居

土家族人民多居住在湘、鄂、川、黔四省毗邻的内陆山区。他们在雨多、雾多、湿度大的山区自然条件下，因地制宜地建造了一种干阑式木构架吊脚楼民居，人住楼上，下养牲畜，以避免住所潮湿，还能防止暴雨时山洪水害。

民居布局，随地形和功能的需要而灵活处理，或依山，或傍水，或向阳，集居在一起，与自然环境协调统一，融人工美与自然美为一体。建筑则视山坡的陡缓，分层筑台，在台地上建房。屋脊既有平行山坡等高线做法，也有垂直等高线的。屋顶有做成同一高度的，也有将屋顶逐级下降而成台阶状的。在陡峭的崖壁或山溪旁等复杂地形建房，往往采取悬挑的方法，以争取使用空间。

灵巧多姿的建筑造型，形成了土家族民居的独特风格。在屋檐处理上，吊脚楼采用了屋角反翘和屋面举折的结构，其外形给人以舒展向上的美感。

Unit 8　Vernacular Houses in the Region of Rivers and Lakes（水乡民居）

【参考答案】

Text A

Task 1

1. Answer the following questions with the information contained in Text A.

1）The layout of the region of rivers and lakes consists of two main types：one type is the village that mainly functions as the residence while the other is the complex that functions both as a trading center and a residence.

2) In a village, a bridge is a necessary transport component to link different parts over the stream-way there.

3) One sees from a distance, the adjoining black-paint gates, the homesteads with grey tiles and white wall situated along the slab-stone paths, along with the poplar trees, the willows trees, the fruit trees and the bending weeping willows, the stretch of wharfs behind them, several small boats drifting on the water of the stream.

4) It is really like an easy, comfortable, peaceful and secluded picture in which people in the region of rivers and lakes can live a life of leisure.

Task 2

2. Fill in the blanks with the given words. You may not use any of the words in the bank more than once. Change the form of the given words if necessary.

1) adjoining　2) boundless　3) consist of　4) complex
5) indispensable　6) landscape　7) layout　8) secluded

Task 3

3. Please match the pictures below with their corresponding bridge type.

1) B　2) D　3) E　4) C　5) F　6) A

4. Translate the following paragraph into Chinese.

远望水乡岸边，毗连的黑漆大门、灰瓦白墙的宅第住家，坐落在石板小路旁，河畔种植着杨柳、果树，弯弯的垂柳，其后一片埠头，几叶小舟荡漾在小河水面，宛如一幅水乡居民享受安逸生活的幽静画面。

【参考译文】

Text A　水乡民居简介

我国南方地区，江湖横贯，溪流密布，众多乡村墟镇常沿河傍水而建。水乡村镇布局中，一类是乡村，以住为主；另一类是商住综合，其特点是河水贯通村内，也有沿村或环村而过。墟镇中常有两条河道汇合，增加了水乡的交通和商业。有些店铺就设在沿河边的通街上，从水陆两路进店都可购物。

墟镇街巷，老百姓的住房——民居建筑，宁静而有序地整齐排列着，屋前有道路，屋后有水路。水陆交通方便，出门即可登船是水乡的一大特色。而船也就成为水乡住户不可缺少的交通工具。乘着小船在水巷中悠悠绕行，足以体会和眷恋水乡生活的无限乐趣和情感。

在村镇中，桥是水乡河道中联系各地各角不可缺少的交通元件。它纵横相连，像蜘蛛织网，成为水乡村镇的美好景观。如直桥、平桥、拱廊、廊桥、圆孔桥、梯级桥，还有宝带桥等各式各式各样的桥梁，为水乡、村镇增添了无限的美好景观。此外，还在桥头建造了一些楼阁或休息亭榭，有的在河边建起了商铺食肆，组成了水乡建筑的一片繁荣景象。

远望水乡岸边，毗连的黑漆大门、灰瓦白墙的宅第住家，坐落在石板小路旁，河畔种植着杨柳、果树，弯弯的垂柳，其后一片埠头，几叶小舟荡漾在小河水面，宛如一幅水乡居民享受安逸生活的幽静画面。

Text B　江苏吴江同里水乡

同里镇位于江苏吴江区东北。镇内风景优美，镇外四面环水。镇区被15条小河分隔成7个小岛，而49座古桥又将小岛串为一个整体。镇内街巷逶迤，河道纵横，家家临水，户户通舟，巷内深邃，幽静宜人，屋瓦连绵，白墙花窗，具有独特的水乡风貌。

同里的优美环境、便利交通和丰富物产成为士绅退隐颐养之地，因此镇上多深宅大院和精良民居。全镇现存明代建筑十余处；清代建筑如退思园等数十处，被称为明清建筑博物馆。

退思园系清代一座十分精致而别具匠心的私家花园。其布局自西往东，为宅、庭、园，横向展开。退思园于江南园林中具贴水园之特例。山、亭、馆、廊、轩、榭皆紧贴水面，园如浮水上。水是万物生命之源，临水、跨水、与水亲近，这是人们生活中最惬意的享受。

Unit 9　Cave Dwellings
（窑洞民居）

【参考答案】

Text A

Task 1

1. Answer the following questions with the information contained in Text A.

1）In the Central Plains of China including Gansu, Shaanxi, Shanxi, and Henan Provinces.

2）The cave dwellings can be classified into three categories: hill-backed cave dwellings, small-courtyard cave dwellings and covering caves.

3）It leaves people a sense of natural beauty in the countryside with simplicity and roughness.

Task 2

2. Fill in the blanks with the given words. You may not use any of the words in the bank

more than once. Change the form of the given words if necessary.

1) shear 2) viscosity 3) stuck on 4) ingredient 5) essence
6) silt 7) averted 8) precipice 9) adjoining 10) ecotope

Task 3

4. Translate the following paragraph into Chinese.

我国中原黄河流域地跨甘肃、陕西、山西、河南等省，有得天独厚的黄土资源。它地质均匀，分布连续，土层厚达 50 ～ 200 米。由于它的主要成分为石英构成的粉砂，颗粒较细，黏度较高，黏聚力和抗剪度较强，便于挖掘施工。窑洞民居既防寒，又保暖，是当地老百姓就地取材，因材致用最合适的居住方式之一。

【参考译文】

Text A 窑洞民居介绍

我国中原黄河流域地跨甘肃、陕西、山西、河南等省，有得天独厚的黄土资源。它地质均匀，分布连续，土层厚达 50 ～ 200 米。由于它的主要成分为石英构成的粉砂，颗粒较细，黏度较高，黏聚力和抗剪度较强，便于挖掘施工，窑洞民居既防寒，又保暖，是当地老百姓就地取材、因材致用最合适的居住方式之一。

窑洞民居根据地形土质分为三类，第一类是直接依山靠崖挖掘横洞成窑，称为靠山窑。再细分又可分两种，一是靠山式，窑前比较开阔，民居沿坡开挖，通常底层屋面是上层窑居的平台。另一种称为沿沟式，窑洞分布在沿沟两岸的崖壁土层。由于沟谷较窄，窑前不开阔，但因沟窄可使两岸窑洞避风沙。

第二类下沉式窑，也称天井窑，在没有条件作靠崖窑的平坦地带，只能就地下挖成四壁闭合的下沉院，然后再向四壁挖窑，称为天井窑。河南称为“天井院”，甘肃称“洞子院”，山西称为“地窑院”“地坑院”。居民出入靠梯道。

第三类覆土窑，也称独立式窑洞，它实质上是一种以土坯或砖石建造的拱形房屋，上部覆土夯实。

窑洞单体平面为带拱券顶的长方形房间，相邻两窑洞中间可打通。窑洞正面为出入门口，称为窑脸，是窑洞的主要正面。下为槛窗，上为拱形窗户，其旁设出入大门。也有的窑洞开窗不开门，通常作为坑床。窑洞民居外观简洁朴实，一般不显露，有的在门框、门楣做一些装饰，有的在窗户玻璃加贴剪纸。它与黄土大地融合在一起，保持自然生态环境风貌，给人一种淳朴、粗犷的乡村自然之美。

Text B 河南巩义康百万庄园

康百万庄园面对伊洛河，背靠邙山岭，建筑坐北朝南，环境幽美。它始建于清道光年间，至宣统年间完工，历时数十年。整个庄园包括祠堂、金谷寨主宅院、普通宅院、作坊、栈房等，是我国北方黄土高原区典型的城堡窑洞庭院住宅。

金谷寨主宅院共4个并列的四合院和1个偏窑崖院组成，除第一院设有堂屋正房外，其他院落都以砖砌崖窑作为正房。头院内房屋纵轴对称、错落有致。第二、三、四院都设有垂花卷棚二门，二门前有4米宽的东西走道相通，后院还有2米宽的通道横贯，形成一个既相互联系又相互独立的大型宅第。

康百万庄园富丽豪华，砖木结构精巧别致，叠脊山墙气宇轩昂，门窗棂花剔透玲珑，雕梁画栋风雅华贵，家具陈设典雅古朴，给庄园增添了浓郁的地方特色。

Unit 10 Garden Houses
（园林民居）

【参考答案】

Text A

Task 1

1. Answer the following questions with the information contained in Text A.

1）The garden houses can be classified into three types: a residence with a study built beside, an independent garden with a style of its own, and one homestead of the residence, the study and the garden grouped together.

2）The means of the assembled view, the scenic focal point and the borrowed view are adopted so as to create more and richer sceneries to satisfy the garden owners' needs to enjoy the sight of the garden, to live in the garden and to hold parties or games in the garden.

3）They used to be inhabited by old and well-known families, literati and officialdom.

Task 2

2. Fill in the blanks with the given words. You may not use any of the words in the bank more than once. Change the form of the given words if necessary.

1）afforest 2）subsidiary 3）odd 4）idly
5）Savor 6）bank up 7）pavilion 8）ornamental

Task 3

3. The following is a list of terms. After reading it, you are required to find the items equivalent to those given in Chinese in the list below. Then you should put the corresponding letters in brackets.

1）F 2）O 3）P 4）D 5）M
6）C 7）J 8）A 9）H 10）E

4. Translate the following paragraph into Chinese.

汉族中有一种特殊类型的民居是供世家、文人、士大夫所用，到后代在一般士民中也有采用，这就是住宅、书斋、庭园结合在一起的民居建筑，称为园林民居，或庭园民居。

【参考译文】

Text A　园林民居介绍

汉族中有一种特殊类型的民居是供世家、文人、士大夫所用，到后代在一般士民中也有采用，这就是住宅、书斋、庭园结合在一起的民居建筑，称为园林民居，或庭园民居。它通常又分为如下三类。

一是宅旁设书斋，也有单独设立者，自成一种类型，称为独立式书斋。它的平面特征是，在三开间民居中，将厅堂向前延伸为长厅，在长厅突出部分左右两旁各设一个小天井，天井内种植纤细的竹木或堆置少量奇石异草，使书斋左右都有庭院绿化。长厅两边都开有较大的窗户，这种书斋虽小，但环境清静幽雅。

二是宅旁设园林，园林独立设置，自成一种类型，它在江南、岭南较多见。其布局特征是，园内不设宅居，而以观赏为主。宅与园连接，有门相通，但各自独立。

三是住宅、书斋、庭园组合在一起，三者功能既有区分，又各自独立。它的平面特征是，以住宅为主，书斋、庭园为辅。宅内有宁静的环境，书斋内可进行词诗书画朗读品赏，闲时则漫步庭园。

园林的设计，以厅馆为中心，四周辅以山石、池水、花木、草卉，并用廊、墙、桥、亭为间隔或相连组成各个景区。它要求在占地不大的有限面积内，充分利用周围的环境、建筑和自然景色，并采用组景、对景、借景等手法，创造出更多、更丰富的景色来达到满足园林主人观赏、可居、可游的目的。

园林民居着重动静结合，以幽雅为主。明计成《园冶》一书指出“三分匠、七分主人”，说明园主是创作园林的主导者，又指出“巧于因借，精在体宜”，这是园林创作设计的准则。通过园林民居的营建实践，不但可以获得宜居宜游的艺术享受，又可获得园林民居中的美学欣赏价值。

Text B　江苏苏州民居园林

苏州私家园林盛于明清。园林多为住宅的一部分，规模不大，其灵活多变的园林空间处理手法、奇绝精妙的叠石理水、恰到好处的植物配置、精巧典雅的建筑群体、淳厚丰富的文化内涵，令人回味不绝。

无水不成园，水是园林的血脉，给园林以无限生机。造园者因地制宜，掘地成池，堆土成山，山水之间高低错落，层次分明。

水令人远，石令人古。山是园林的骨骼，叠石堆土，仿真山之脉络气势，做出峰峦丘壑、洞府峭壁，使园景变化万千。园林叠山选料，以太湖石为主，上好的太湖石兼具皱瘦漏透四大特点：皱，即指石块有波折、有层次、有变化；瘦，指石的挺拔清秀；漏，使石更富于情趣；透，即石峰玲珑剔透。

苏州园林建筑汲取江南民间建筑的精华，轻巧纤细、玲珑剔透，其室内外空间通透，外观粉墙黛瓦，园内山水花木，显示出一种恬淡雅致的风貌。